PENGUIN B[OOKS]

D0453933

A FIELD GUIDE TO RADIATION

Wayne Biddle won a Pulitzer Prize for his reporting on the "Star Wars" antimissile project. He is the author of five previous books, including *A Field Guide to Germs*, winner of the American Medical Writers Association's Walter C. Alvarez Honor Award, and *Dark Side of the Moon*, which was selected as a *New York Times Book Review* Editor's Choice. He teaches at Johns Hopkins University.

ALSO BY WAYNE BIDDLE

Dark Side of the Moon

A Field Guide to the Invisible

A Field Guide to Germs

Barons of the Sky

Coming to Terms

A Field Guide to
RADIATION

Wayne Biddle

 PENGUIN BOOKS

PENGUIN BOOKS

Published by the Penguin Group

Penguin Group (USA) Inc., 375 Hudson Street, New York, New York 10014, U.S.A.

Penguin Group (Canada), 90 Eglinton Avenue East, Suite 700, Toronto,
Ontario, Canada M4P 2Y3 (a division of Pearson Penguin Canada Inc.)

Penguin Books Ltd, 80 Strand, London WC2R 0RL, England

Penguin Ireland, 25 St Stephen's Green,
Dublin 2, Ireland (a division of Penguin Books Ltd)

Penguin Group (Australia), 250 Camberwell Road, Camberwell,
Victoria 3124, Australia (a division of Pearson Australia Group Pty Ltd)

Penguin Books India Pvt Ltd, 11 Community Centre,
Panchsheel Park, New Delhi - 110 017, India

Penguin Group (NZ), 67 Apollo Drive, Rosedale, Auckland 0632,
New Zealand (a division of Pearson New Zealand Ltd)

Penguin Books (South Africa) (Pty) Ltd, 24 Sturdee Avenue,
Rosebank, Johannesburg 2196, South Africa

Penguin Books Ltd, Registered Offices:
80 Strand, London WC2R 0RL, England

First published in Penguin Books 2012

10 9 8 7 6 5 4 3 2 1

Pages 257–258 constitute an extension of this copyright page.

LIBRARY OF CONGRESS CATALOGING IN PUBLICATION DATA
Biddle, Wayne.
    A field guide to radiation / Wayne Biddle.
       p. cm.
    Includes bibliographical references and index.
    ISBN 978-0-14-312127-5
  1. Radiation.   2. Radioisotopes.   3. Radioactive substances.   I. Title.
    QC475.B484 2012
    539.2—dc23   2012009025

Printed in the United States of America
Set in Adobe Garamond Pro
Designed by Judith Stagnitto Abbate / www.abbatedesign.com

MIDORI NAKA

JUNE 19, 1909–AUGUST 24, 1945

無

Contents

CONTENTS

CONTENTS

CONTENTS

CONTENTS

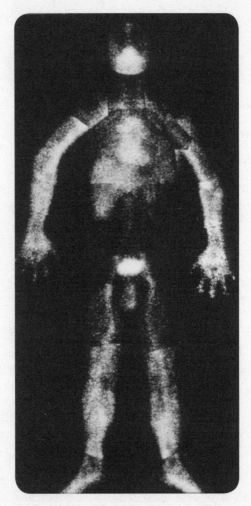

Do you feel like I do? A composite photograph showing the movement of isotopes in the human body.

Introduction

SINCE THE MANHATTAN PROJECT made the energy that holds matter together available as a useful source of power, there have been five occasions when life has been threatened on a wide scale by atomic devices: the bombings of Hiroshima and Nagasaki in 1945; the outdoor testing of nuclear weapons in many locations from 1945 until 1980; the reactor meltdown at Three Mile Island in Pennsylvania in 1979; the reactor explosion at Chernobyl in Ukraine in 1986; and the multiple reactor failures at Fukushima in Japan in 2011. Although radiation is a fundamental process of the universe, these events of the last seven decades represent together the first time in 3.8 billion years that radiation has harmed life in an acute way.

We live under a kind of duress for which we did not evolve. And we are trapped, because the bombs and reactors we have built cannot be completely removed now. Even if the construction of new ones were halted and the old structures were destroyed (the Japanese government estimates that decommissioning the ruined Fukushima Daiichi plant will take forty years and require robots and other new technologies that do not yet exist), the huge quantities of radioactive material that fueled them will remain dangerous for tens of thousands of years. And we have no safe place to put it all—just defining what "safe" could mean for that much time seems absurd. Nuclear technology has the alarmingly unique characteristic that intentional or accidental release of massive amounts of radiation from this fuel in a small locality quickly spreads around the globe, creating hazards far away. In effect, our planet is newly hostile, and we need to find a practical way to mitigate the risk of living on it.

Unfortunately, radiation is not an intuitively obvious subject like, say, the weather. After its discovery during the last decade of the nineteenth century, radiation took some of the most inspired scientific research in modern history to decipher its nuances. All elements have at least one radioactive isotope, but there are dozens whose radioisotopes are not found in nature; they must be produced artificially. All elements with 83 or more protons in their nucleus are radioactive in all forms, but so are technetium, with 43, and promethium, with 61. Fortunately, nonexperts can acquire a working knowledge of radiation basics, just as they can learn how to sort out the vast intricacy of the earth's flora and fauna. Field guides were developed long ago for this general purpose. Carl Linnaeus (1707–1778), the great Swedish botanist and father of modern taxonomy, quipped in 1737 that the difference between an animal and a botanist is that a botanist is equipped to communicate knowledge about plants to others. The first book of French naturalist Jean-Baptiste de Monet de Lamarck (1744–1829), published in 1779 and called *Flore françoise*, described some four thousand species of plants. Its 1,200 copies sold out quickly and, most important, it was reprinted after the French Revolution, when the formation of democratic *Écoles centrales* fueled the publication of innovative textbooks. (Lamarck no doubt aided his career, and perhaps saved his head, when, as keeper of Louis XVI's herbarium in 1790, he changed the name of the royal Jardin du Roi to Jardin des Plantes.) Laymen appreciated that Lamarck wrote in common French, rather than Latin. A second edition of *Flore*, in 1805, sold five thousand copies, an extraordinary number for the era. The birth of English-language field guides had to wait several decades for the dilution of post-Revolution anti-French sentiment among the British intelligentsia. Thomas Nuttall's *Manual of the Ornithology of the United States and of Canada* (1832–1834) is considered the first such resource. The shift from buckshot to binoculars ensured a mass market for bird guides in the twentieth century, and today they are read by people who may never look beyond their kitchen window, serving as a model for organizing other complex subjects.

Field guides are not scholarly texts. They eschew jargon and are utilitarian handbooks for amateurs and professionals that help to identify and explain something intricate or hard to find. Because ionizing radiation cannot be detected by our senses, we are engulfed by energy that we cannot locate at all without specialized instruments, which, while readily available, are not common consumer gadgets. Anyone living within fifty miles of a nuclear power plant might want to own some pocket dosimeters, but most people will continue to rely on professional measurements communicated through government offices or news media, for better or worse. Likewise, anyone facing a series of medical X-rays needs to ask the doctor about management of exposure. A field guide to radiation is not pro- or anti-radiation any more or less than a field guide to reptiles is pro- or anti-snake. The reader can take the book in hand, apply its contents against what is found in nature or the news, and gain some reliable orientation. Where there are implications for health and wellness, they are sorted out with the goal of clarifying choices for individual action. One envisions the concerned naturalist sitting on a log regarding the snake, reading about its mysterious ways, and then deciding how to coexist with it.

What does coexisting with radiation mean? Keeping away from it, if possible. Although there are still charlatans afoot who say that small doses are beneficial, and mainstream scientists who accept the status quo of livable exposures, the basic postulate followed here is that there is no completely safe dose, even for natural "background" radiation and medical diagnostic or therapeutic procedures. This is the scientific consensus expressed by the 2005 BEIR (Biological Effects of Ionizing Radiation) VII report from the National Academy of Sciences, which found that risk rises with exposure at any level above zero.

Deciding how much risk is acceptable is a contentious

social and political issue, one that shows no sign of abating as nations struggle with how to power the mechanical necessities of life. For more than two hundred years, societies with any wherewithal have built economies that depend on copious supplies of artificial energy. A world of prudently conserved renewable energy is so far in the future that no government has taken a decisive step beyond the centralized generation of electricity from fossil fuels or enriched uranium. Quite the contrary—advanced industrial countries such as France and Japan have hitched their wagons to the star of nuclear power. No disaster or long-term prediction has yet made a difference. After Fukushima, Germany declared that it would abandon nuclear power—which provided 22 percent of its electricity—and close its eighteen reactors by 2022, filling 80 percent of its needs with energy from wind and solar generators by 2050, but it remains to be seen how much reliance will fall on new coal and gas plants. Meanwhile, without importing massive amounts of electricity, it is threatened by blackouts. Finding radioactive hotspots in Tokyo months after a reactor meltdown 160 miles away was not enough to budge the Japanese government. "Nobody stands in one spot all day," responded Kaoru Noguchi, head of Tokyo's health and safety agency. "And nobody eats dirt."

Such bureaucratic callousness is reminiscent of the cold war bravura of T. K. Jones, an undersecretary of defense in the Reagan administration, who declared in 1981 that a nation can survive nuclear war "if there are enough shovels to go around." (The shovels would be for digging holes as fallout shelters.) "It's the dirt that does it," Jones explained about shielding the populace from radiation. Nuclear war remains hypothetical, but when a nation of 127 million people with only 146,000 square miles of territory—Texas has 25 million dwelling on 267,000 square miles—faces the long-term loss of *any* living space due to fallout from a reactor disaster, it is perhaps not surprising that its officials would feel pressure to skirt scientific reality. At least 400 square miles of land must be decontaminated around Fukushima, which will create some 40 million cubic yards of soil and debris that no one knows how to dispose of safely. For many of the 160,000 people displaced from the region, returning home will not be possible during their lifetime. At an elementary school in Naraha, a town that had been evacuated after the March 2011 Fukushima Daiichi debacle but that the government declared safe for residents to return to in September, radiation was measured at 0.77 microsieverts per hour, or 6.75 millisieverts a year—far above the annual limit of 1 millisievert for civilians that is recommended by the International Commission on Radiological Protection. And children and pregnant women are most vulnerable to radiation. About 3 percent of Japan's landmass is now contaminated above the safety standard.

When vital technology fails, political necessity fills the void. It is in this morass that accurate facts become scarce as entrenched interests vie for the upper hand, leaving ordinary citizens to fend for themselves. Radiation Defense Project, the Tokyo grassroots organization that discovered concentrations of cesium radioisotopes in that city comparable to levels inside protection or relocation zones around the devastated Chernobyl reactor, demonstrated the inevitable loss of faith in central authority. "If the government is not serious about finding out," said one of its leaders, "how can we trust them?" As of February 2012, only three of Japan's fifty-four nuclear reactors (which had generated 30 percent of the nation's electricity) were operating, with the rest likely to be turned off soon, because local communities blocked restarts after regular maintenance shutdowns. Perhaps cultural disintegration is the most insidious consequence of nuclear disaster, an effect familiar to historians of the Black Death that depopulated Europe during the fourteenth century and ripped apart the social contract wherever it struck.

WITH THE CLEAR LENS of hindsight, it is obvious that there was something irresistibly magical about radiation right from the start. The pioneer researchers of the 1890s—Wilhelm Roentgen in Germany, Marie and Pierre Curie in France—were mesmerized by it even as they quickly realized that the invisible rays could ulcerate their skin. "One of our joys," Marie Curie wrote, "was to go into our workroom at night; we then perceived on all sides the feebly luminous silhouettes of the bottles or capsules containing our products [i.e., radium]. It was really a lovely sight and always new to us. The glowing tubes looked like fairy lights." Within several years, a laboratory curiosity became a public sensation, with Thomas Edison rushing an X-ray machine from New Jersey to Buffalo, New York, in 1901 to help locate an assassin's bullet in President William McKinley. (The machine was not put to use, and McKinley died from sepsis and gangrene a week after being shot.) Well into the 1930s, gullible consumers fell prey to quack elixirs laced with radium, even as Marie Curie herself died from leukemia after decades of poor health clearly caused by radiation exposure. For his part, Edison abandoned the "fluoroscope" as a commercial venture in 1904 after the death of his technician, Clarence Dally, from a degenerative skin disease that progressed into a carcinoma. "The x-rays had affected poisonously my assistant," Edison said.

Of course, if not for the Manhattan Project of World War II, radiation might have remained a medical tool of increasing sophistication and manageable risk. The discovery of nuclear fission in 1939, just before Hitler invaded Poland, was a darksome coincidence that first spawned a nightmarish weapon and then a troublesome source of energy—one championed, out of guilt or hope, by the same scientists who had built the Bomb. Hans Bethe (1906–2005), head of the Manhattan Project's theoretical division, spent the rest of his long life advocating nuclear power. This cultural synergy between nuclear weapons and nuclear reactors, twisted by cold war paranoia, thrived in a climate of official secrecy and deception that soon made it impossible to draw clear lines between military, diplomatic, and commercial ambitions. In the United States, the Atomic Energy Commission systematically lied to the public about fallout from nuclear weapons tests in the open atmosphere, putting higher priority on perfecting the weapons than on protecting public health. The American diplomatic campaign known as "Atoms for Peace" helped to export the reactor designs of widely admired corporations such as General Electric to countries hungry for postwar economic development, such as Japan. The United States still has some twenty-seven nuclear cooperation agreements with foreign countries and international organizations, under which reactor fuel and hardware are sent abroad. Thousands of kilograms of highly enriched uranium and tens of thousands of kilograms of plutonium have accumulated overseas—so much that a comprehensive inventory continues to elude federal auditors. Hundreds of tons of this radiotoxic, weapon-usable material are simply unaccounted for. New American contracts may or may not contain provisions against uranium enrichment and plutonium reprocessing by client states, but French and Russian exporters make no such demands.

Given this boundless landscape of man-made radiation, it would seem futile to expect the genie to return to the brazen vessel, to borrow Walt Disney's imagery from his popular 1957 propaganda film *Our Friend the Atom*. Once released, great technologies disappear only through obsolescence, as they are replaced by something more efficient or more rewarding or more pleasurable. It seems reasonable to posit that there is nothing on the horizon that would make

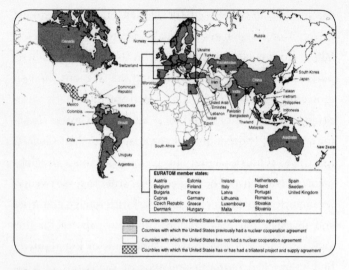

U.S. Nuclear Cooperation Agreement partners

either nuclear weapons or nuclear power obsolete, besides a fundamental reordering of industrial civilization. History offers examples of catastrophes that have led to such change, but not of rational planning. The United States still deploys about 2,500 nuclear weapons and stores 2,600 more. And nations still take great risks to join the nuclear club.

Essentially, we are still living in the technological landscape created by World War II and the ensuing cold war. The cultivation of science by government, which was pioneered in Germany in the late nineteenth and early twentieth centuries with heavy emphasis on military needs and economic development, served as a model for the Manhattan Project and subsequent cooperation among governments, corporations, and universities that produced the nuclear world. The Soviet Union accomplished many of the same objectives with a somewhat different model, but it is now gone from the scene—though its nuclear detritus remains very much in play. A century ago, when Henry Adams chose the dynamo to symbolize the kind of power that would soon displace spirituality—or the Virgin—as the strongest organizing factor in society, he was off by only one step. The dynamo needs to be spun, and how to do that turned out to be the master force. The world is now organized to keep the dynamos spinning, so much so that calling into question any aspect of that organization provokes accusations of being anti-progress. Besides global warming, there is no more intractable problem for the planet's future than man-made radiation. How ironic that one leading solution for the former would increase the latter.

THIS FIELD GUIDE contains essays on broad topics, such as medical and occupational radiation, and portraits of numerous radioisotopes—from actinium-227 to zirconium-95—that comprise the natural and artificial universe. There are also explanations of the maddening array of technical measurements—sieverts, becquerels, grays, etc.—used to quantify radiation and exposure to it. Laced into the science is historical commentary, in the hope that this will forestall the tendency of science to look like an activity that takes place outside society. The atom and the bomb and the reactor have occupied the minds of many brilliant men and women. The pioneer physicists and radiochemists, especially, existed in such tight coteries that they often seemed oblivious to the implications of their work. War changed that, of course, but science and engineering and medicine are always isolated from wider circles by technical language and hauteur that exclude nonexperts. This might be tolerable if the laboratory were truly isolated from the street, but ever since Henri Becquerel noticed photographic film being fogged by something shut inside a drawer on a cloudy Paris day, there has been no escape for anyone from radiation above and beyond what the gods intended.

A Field Guide to
RADIATION

Absorbed dose

IN COMMON LANGUAGE, this term is just what it sounds like: the radiation that gets into you. But as dosimetric jargon, it carries important nuances, because not all types of radiation have the same biological effects and not all types of living tissue are equally vulnerable.

The absorbed dose is defined technically as the amount of energy, measured in joules, imparted by any ionizing radiation per unit mass of any material, measured in kilograms. One joule per kilogram (J/kg) of absorbed dose is called one GRAY (Gy). An old unit called the "rad" (for radiation absorbed dose) is still sometimes encountered in industry and government documents, especially in the United States. One gray equals 100 rads.

The absorbed dose is a raw physical quantity and does not indicate much about the risk of biological effects in specific organs or tissues. The high doses used to kill cancerous cells and tumors, in the range of 40 Gy and up, are lethal to all human living matter, so physicians simply use the gray when ordering radiotherapy doses. (So-called fractionated doses—large totals delivered in small daily amounts—are used to limit side effects, but any cancer patient knows the debilitating nature of radiotherapy, the word *therapy* here being a special kind of euphemism for a process akin to turning off a lightbulb with a sledgehammer.) The spectrum of health effects from whole-body exposure as doses rise is also usually expressed in grays (see ACUTE RADIATION SYNDROME). When more finely tuned information is needed about the effects of different kinds of radiation on different kinds of tissues and about the long-term risk of damage, a unit called the SIEVERT is used.

Actinium-225, -227

ACTINIUM IS A RARE ELEMENT of little practical use, but worth keeping an eye on because it happens to be a DECAY PRODUCT of two of the nuclear world's great denizens, PLUTONIUM-239 and URANIUM-235. There are dozens of actinium RADIOISOTOPES, but only actinium-227, with a HALF-LIFE of 27.7 years, occurs in notable amounts. A gram of natural uranium contains perhaps 2×10^{-10} gram of actinium, but the ENRICHED URANIUM used in reactors and weapons contains more, making actinium-227—a BETA PARTICLE emitter, albeit weak—of potential environmental concern. Almost all of the radiation associated with actinium-227, which was once considered to be "rayless," comes from *its* decay products. Actinium-225 is used in nuclear medicine to generate its decay product, BISMUTH-213, whose ALPHA PARTICLES kill tumor cells such as leukemia.

At the suggestion of Pierre and Marie CURIE, the French chemist André-Louis Debierne (1874–1949)—a shy oddball who had helped the Curies refine pitchblende ore during their discovery of POLONIUM and RADIUM in 1898—looked for additional radioactive elements that might still be hiding in the ore. In 1899 he claimed to have found one, which he named actinium after the Greek word *aktis*, for "ray," in keeping with the perception of radioactivity at that time. But he might have jumped the gun. Four years later, the pioneer German radiochemist Friedrich Giesel (1852–1927) announced that he had discovered a new element in pitch-blende, which he called emanium, based on the contemporary notion of radioactivity as an emanation. Debierne ignored Giesel's inquiries on the matter, and a face-to-face test in 1904 with the Curies in attendance did not resolve the issue of who discovered what. World War I put an end to Debierne's and Giesel's careers, and both eventually died from the poisonous effects of working with radioactive materials. Actinium and emanium were shown by other scientists to be the same substance, but whether Debierne or Giesel deserves priority credit is still debated by historians. The great Ernest Rutherford threw his weight behind using the name actinium, though he was unimpressed with Debierne's claim to having discovered it, opining that "he had not enough radioactive sense to find it out." Giesel's tombstone in Braunschweig, Germany, bears the inscription "Entdecker des Aktiniums."

Acute radiation syndrome (ARS)

"ATOMIC BOMB DISEASE," the "new sickness," "radiation sickness," "radiation poisoning"—the sad history obscured by an acronym is evident in the colloquial phrases that have been applied over the years to the deadly effects of radiation. Maybe ARS is not as pervasively frightening as AIDS, because not as many people have suffered from it, but behind both cold medical terms are years of denial and suppression.

That radiation is harmful to health was known as soon as man-made radioactivity existed in the 1890s. Early researchers burned their hands with X-RAYS and witnessed the spread of skin cancers. Long before this decade, underground miners in certain regions of Europe were known to die at alarming rates from lung ailments that could not be explained by dust inhalation only (see RADON). The medical and, sometimes, quack entertainment value of X-ray photos and radium-laced consumer products during the first half of the twentieth century meant that firsthand knowledge of the risks of radiation exposure was often ignored by doctors and hucksters alike. After World War II the American government actively concealed knowledge about FALLOUT from the public.

Today, while the detrimental health effects of radiation are no longer denied, government and industry may still act to conceal information or dismiss risky exposures as "safe," as in the months following the Fukushima disaster in Japan. If and when the levels of exposure become known during any such calamity, radiation sickness can be easily diagnosed and, with luck, treated.

ARS applies specifically to when people receive high doses of penetrating radiation over their entire bodies within a short period of time (usually minutes). A range of symptoms then appear, depending on the dose. Extensive information about such events has come from the two atomic bombings at Hiroshima and Nagasaki in 1945, and since then, from industrial accidents such as the 1986 Chernobyl reactor fire in the former Soviet Union. "Large dose" here means greater than 0.7 gray, though relatively mild effects can happen at 0.3 Gy. At this lower end of the exposure scale, symptoms come from destruction of bone marrow and include nausea, vomiting, and fever. With careful treatment, many victims will recover within a few weeks to several years. A dose of 1.2 Gy will kill some patients from infections and internal bleeding; 2.5 to 5.0 Gy will kill half of everyone within sixty days. At the next level, above 10 Gy but sometimes starting around 6, severe diarrhea also occurs as the cells lining the gastrointestinal tract die. At 10 Gy, all

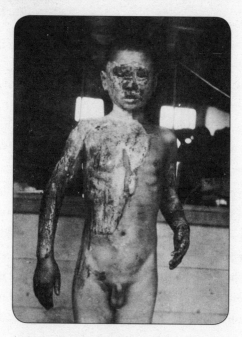

A most acute situation: seventeen-year-old boy
with flash burns, Hiroshima, 1945

exposed patients will die within two weeks. Above 50 Gy,
but sometimes beginning at 20, the cardiovascular and central nervous systems collapse, leading to convulsions, coma,
and death within three days.

At Chernobyl, 134 workers suffered high doses of radiation that caused ARS symptoms that killed 28 of them.

Alpha particles

ONE OF FIVE VARIETIES of radiation that entails serious risks to human health (the others are BETA PARTICLES, GAMMA RAYS, NEUTRONS, and X-RAYS). These are known as *ionizing* radiation, because they have enough strength to disrupt how matter, including living tissue, holds together. (*Nonionizing* radiation, such as visible light, radio waves, and the microwaves in kitchen ovens and in CELL PHONES, does not pack enough punch to do this, though it can cause other biological effects of lesser concern.) Alpha particles are generally harmless to people unless the material that emits them somehow gets inside us. Then they are potent causes of cancer.

During the 1890s and 1900s, when scientists in Germany, France, and England first described various kinds of radioactivity, they referred to the strange new radiant energy as "rays"—*Strahlen* in German, *rayons* in French—using the common words for a beam of light. Unlike light, these rays had miraculous penetrating power, which Henri BECQUEREL first noticed during a cloudy February in Paris in 1896. After putting into a laboratory drawer an experiment with uranium salt crystals (potassium uranyl sulfate) that needed sunshine to stimulate luminescence—the ability of some substances to glow in the dark after being exposed to light—he found, several days later, that a nearby photographic plate held a blotchy image. Evidently something was shining out of the uranium salts *through the wooden drawer*, even without the sun's help. After feverish study, he christened the phenomenon *les rayons uranique*, showing that they streamed continuously out of uranium. He had, in fact, discovered radioactivity, for which he shared the 1903 Nobel Prize with two other pioneers, Pierre and Marie CURIE (who coined the term).

News of Becquerel's work traveled quickly across the Channel, to the Cavendish Laboratory at Cambridge University, where a young New Zealander named Ernest Rutherford, who is unquestionably the father of the nuclear age, was beginning what would stretch into nearly ten years of radioactivity research. His first important find, in 1898, was that uranium rays had two components. One, which he called alpha radiation, was positively charged and could be completely absorbed by a sheet of paper or aluminum foil as thin as eight ten-thousandths of an inch. The second, dubbed beta radiation, was negative and a hundred times more penetrating.

The truly wondrous thing about alpha rays was that when an element emitted them, it turned spontaneously into a different element. Uranium, for example, changed into thorium. "Don't call it *transmutation*," Rutherford facetiously warned a colleague; "they'll have our heads off as alchemists."

He concentrated on alpha rays, concluding in 1908—the same year he won the Nobel Prize—that they were beams of particles like charged helium atoms, which Marie Curie had foreseen when she opined that Becquerel's uranium rays might be due to changes within atoms rather than to familiar chemical reactions. Whatever caused alpha particles to pop out of uranium released a million times more energy than any chemical process. It took another two years to identify them as helium ions (clusters of two protons and two neutrons, a configuration not understood at the time), actually atomic nuclei, a totally new concept in atomic structure. As soon as an alpha particle lost most of its energy after traveling a short distance, it stole two electrons from its surroundings and became a harmless helium atom.

Henri Becquerel's photographic plate fogged by alpha particles from uranium salt crystals. A Maltese Cross blocks the particles in the lower image, leaving a shadow.

But that theft can be lethal. Alpha particles in nature usually do not hurt us, because the first layer of dead cells on unbroken skin is too thick for them to penetrate. Nor can they make it to the lens of the eye. They certainly cannot get through clothing. In fact, very few alpha particles can travel more than several inches through air before losing their momentum. (The exceptions are the extremely high-energy alpha particles contained in COSMIC RAYS.) If we inhale or ingest the elements that emit alpha particles, however, or if they enter through skin wounds, they can be horribly damaging as they career through soft tissues, stealing electrons and destroying cellular molecules, thus killing the cells. Under these circumstances, they are the most destructive form of ionizing radiation. A dose of alpha radiation delivered to living tissue this way is twenty times more destructive than the same dose of X-rays. Moreover, a so-called bystander effect has been conclusively demonstrated for alpha particles, whereby irradiated cells or tissues have deleterious effects such as mutations and malignant transformations (from healthy to cancerous) on the nonirradiated. The frequency of malignant transformations when only 10 percent of a group of living cells are pierced by an alpha particle is as great as when all the cells are exposed, a fact that emphasizes the undiminished hazard of alpha radiation at very low doses.

A single alpha particle can cause major genomic changes in a cell. Even allowing for a substantial degree of natural repair, the passage of a single ion has the potential to cause irreparable damage in cells that are not killed. Because there is solid evidence that most cancers are of monoclonal origin—that is, they start from damage to a single cell—these facts provide the basis for assuming a linear relationship between alpha particle dose and cancer risk even at exposure levels where the probability of more than one alpha particle crossing through a cell is very small. At such levels, most cells are never bombarded by even one alpha particle, but there is still some risk of cancer.

As with any particles in the air we breathe, a fraction of inhaled alpha emitters will lodge in our lungs. Depending on how easily they dissolve, they may stay there for a long time or be slowly absorbed into the bloodstream, from where they will collect in the liver or bones. The continuous exposure—sometimes for many years if the emitter has a long HALF-LIFE (such as PLUTONIUM)—of a small region of the body to the short-range but high-strength alpha particles can cause malignant tumors. For example, bone cancer has been linked with exposure to alpha radiation from RADIUM-224. Acute effects are also possible. In 2006 the Russian dissident Alexander Litvinenko was murdered by the injection of about ten micrograms of alpha emitter POLONIUM-210, which killed him in three weeks.

Common alpha emitters include the elements PLUTONIUM, RADON, RADIUM, THORIUM, URANIUM, and AMERICIUM.

Americium-241, -243

AMERICIUM IS A MAN-MADE ELEMENT that found its way from the Manhattan Project of World War II into a ubiquitous home gadget: the smoke alarm. A by-product of plutonium production, it was created secretly in 1944 by a research group led by the nuclear chemist Glenn Seaborg (1912–1999) while they were exploring how to extract PLUTONIUM-239 from uranium for bomb fuel. Americium—patriotically named to match the element europium—would probably have become known only as a component of FALLOUT if not for the popularity of inexpensive smoke detectors that use a tiny piece of one of its nineteen RADIOISOTOPES, americium-241.

Operating on the same principle as a GEIGER COUNTER, a smoke alarm contains an air-filled tube, called an ionization chamber, that conducts the electricity generated when ALPHA PARTICLES from about 0.29 micrograms of americium-241 collide with oxygen and nitrogen. If smoke particles enter the tube, the current is disrupted, which triggers an alarm. Any stray alpha particles are absorbed by the device's plastic case, so it is not considered hazardous to health, unless you like to eat smoke detectors or make bon-fires with them. Weak GAMMA RAYS are given off, too, pro-ducing a dose of about 27 μSv (microSIEVERTS) if you stood stock-still a yard away for a year. Why put a radioactive substance in an everyday consumer item? Because it's the cheapest way to go, of course.

Only several kilograms of americium have to be pro-duced annually to satisfy the demand for smoke detectors. Because old units are usually thrown away like any common household refuse, or are part of the rubble of demolished buildings, millions upon millions of them are accumulating in municipal landfills. With a HALF-LIFE of 432 years, active americium tablets about the size of a pencil eraser will be underfoot in dumps for a long time. Waste, including low-level radioactive waste, is supposed to be buried to prevent its near-term release, but plants and animals can bring it to the surface, where it will be picked up by the wind. The fact that many people disconnect the detectors required by hous-ing codes after a few false alarms raises the question of exactly what benefit is balancing the risk of this radioactive trash.

Americium is also produced by the FISSION reactions in nuclear power plants, which makes it part of the telltale gang of RADIONUCLIDES released into the environment when something goes very wrong. At a boiling water reactor such as those destroyed at Fukushima, spent fuel assemblies—bundles of hundreds of metal rods containing pellets of ceramic uranium dioxide—weighing about 680 kilograms will contain about 220 grams of americium. Indeed, in soil samples taken from a playground near Fukushima, americium-241 was found at concentrations around 0.04 Bq/kg. By comparison, americium-241 from 1940s and '50s nuclear weapons test fallout was found in village surface soil on Bikini Atoll in 1997 at 220 Bq/kg.

Another radioisotope, americium-243, has a half-life of 7,370 years and is also part of the "nuclear fuel cycle." That is, it is produced at various stages of the operation of nuclear reactors. As always, workers in industries where americium is present are at higher risk of exposure than the general public. At the Hanford, Washington, nuclear weapons plant, a sixty-four-year-old worker was massively contaminated in 1976 with americium-241 when a "glove box"—a sealed container with built-in gloves for manipulating dangerous materials—exploded and sent tainted glass shards into his skin. He received intensive treatment and suffered from various blood cell disorders, but died eleven years later of preexisting cardiovascular disease.

Americium is one of only two elements (the other is curium) that are patented—both by Glenn Seaborg.

Antimony-122, -124, -125, -126, -127

THERE ARE DOZENS OF RADIOISOTOPES of the element antimony, but only these five have a significant HALF-LIFE. Antimony-125's is longest, at about 2.75 years, with antimony-124's next, at about two months. The half-lives of the other three are measured in days. In commercial nuclear power reactors, nonradioactive antimony is often combined in an alloy with BERYLLIUM to provide a dependable source of neutrons for FISSION reactions in the core. Strong GAMMA RAYS from antimony-124, which is generated in the alloy during normal operation by getting bombarded with fission neutrons from the core, stimulate the production of more neutrons from the beryllium. Even if the reactor is shut down for several months, this source can be used to start it up again, rather like a charged battery in a car that has been parked awhile. Like all the other hazardous materials in the plant, radioactive antimony might be released into the biosphere by a meltdown or other serious breach.

The sulfide of nonradioactive antimony (Sb_2S_3) was used in antiquity to make a paste for eye cosmetics, or kohl. As the natural mineral called stibnite, it was also valued as an eye medication. The prophet Muhammad is thought to have recommended it for clearing vision and making hair sprout. Shall we note here that this would not be the effect of radioactive antimony?

Argon-37

THE RADIOISOTOPE argon-37 is a definitive telltale of a subterranean nuclear explosion, created when neutrons from the blast strike calcium in the surrounding soil (specifically calcium-40, which makes up 97 percent of all natural calcium). On-site measurement of argon-37 gas seeping out of the ground is one way of verifying the Comprehensive Nuclear Test Ban Treaty, which was adopted by the United Nations in 1996 to halt all types of nuclear weapons detonations (and remains unratified by China, India, Israel, Pakistan, the United States, and several other nuclear states).

Since the 1950s, when public fear mounted about FALL-OUT from bomb tests in the open atmosphere, various international agreements have tried to end the explosions. The Partial Test Ban Treaty of 1963 and the Nuclear Non-Proliferation Treaty of 1968 effectively halted the cold war spree of atmospheric testing by the United States and the Soviet Union, but other nations continued on a more limited scale while the two superpowers took their weapons programs underground. Political opponents of arms-control treaties have always tried to raise doubts about whether they can be reliably enforced and verified. Radionuclide monitoring is just one of many techniques, including seismology and acoustics, that today make clandestine tests anywhere around the planet virtually impossible.

Background radiation

THIS TERM, which evokes the innocuousness of Muzak, is also used in astronomy and astrophysics to describe the afterglow of electromagnetic (microwave) energy left in space from the primordial Big Bang that created the universe fifteen billion years ago. Like the hiss behind a tape-recorded symphony, the faint whisper is there no matter where you listen. In the context of ionizing radiation, however, background radiation means something more worrisome. It has multiple sources, varies from place to place, and is not entirely harmless, despite being ubiquitous.

Perhaps no other bit of nuke-speak has been used more insidiously to palliate public fears about radiation. The U.S. Nuclear Regulatory Commission (NRC), which Barack Obama has criticized for being "captive of the industries that it regulates," still leads its "Fact Sheet on Biological Effects of Radiation" by reassuring readers that "radiation is all around us" and "our nation's Capitol, which is largely constructed of granite, contains higher levels of natural radiation than most homes." These are not exactly lies, but neither is calling a wolf a big dog and leaving it at that.

For many people around the world, exposure to natural radiation is greater than exposure to man-made sources. In the United States, natural background radiation accounts for 82 percent of the population's annual exposure. It is the permanent, raised foundation upon which all other exposures accumulate. Because there is broad scientific consensus behind the postulate that even the smallest dose of radiation potentially harms human health (the "LINEAR NO-THRESHOLD," or LNT, risk model), seemingly minuscule natural background exposures are not "safe" in the sense of zero risk, an often-heard claim.

The worldwide average annual natural exposure is 2.4 millisIEVERTS (mSv), but this figure does not necessarily cover any particular person, with real-life exposures generally ranging from 1 to 13 mSv. In the United States, the average is about 2.95 mSv; in Japan, 1.5 mSv. For the sake of rough comparison, a typical chest X-RAY is about 0.06 mSv. About 25 percent of the earth's population get less than 1 mSv, 65 percent get between 1 and 3 mSv, and 10 percent get more than 3 mSv. Large areas of the world receive as much as 16.9 mSv, and even higher values occur locally. Wherever you are, this is what you absorb 24/7 for a lifetime, pretty much no matter what.

Humans evolved with natural radiation, of course—it no doubt caused some of the random genetic mutations that make us *us*—and must continue to live with it to varying degrees. For example, COSMIC RAYS from outer space are not a lifestyle choice. They originate in exploding stars and, outside the earth's atmosphere, consist mostly of protons and ALPHA PARTICLES. When they enter the atmosphere, they collide with the nuclei of various atoms in the air and generate more ionizing radiation particles. Exposure varies with latitude and altitude above sea level, as the earth's atmosphere and magnetic field shield some of them out, but they are everywhere. (Residents of La Paz, Bolivia, who live at 12,795 feet, receive about 2.02 mSv per year, of which 40 to 50 percent is from NEUTRONS, which are the high-LET [for linear energy transfer—see GAMMA RAYS] component of cosmic rays. In Denver, Colorado, at 5,282 feet, the annual exposure is 570 microsieverts [μSv].) Additionally, soil, rocks, and water contain various primordial RADIOISOTOPES of common elements, such as POTASSIUM-40, URANIUM-238, and THORIUM-232, which in turn spawn RADIUM and RADON through radioactive DECAY. (There is an acronym, of course, for all this stuff: NORM, for "naturally occurring radioactive material.") Plants and animals, including our own wonderful corpus, naturally carry the potassium and carbon isotopes, and thus constantly expose themselves, in a non-salacious sense. You are what you eat. Edward Teller (1908–2003), father of the H-bomb and master purveyor of cold war shibboleths, was fond of suggesting that if you snuggled with two lovers, you would get more radiation via GAMMA RAYS from their potassium-40 than by sleeping against a nuclear reactor. Oh, KoKo.

In most circumstances, these natural sources are too weak

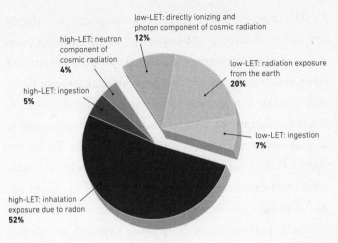

low-LET: directly ionizing and
photon component of cosmic radiation
12%

high-LET: neutron
component of
cosmic radiation
4%

high-LET: ingestion
5%

low-LET: radiation exposure
from the earth
20%

low-LET: ingestion
7%

high-LET: inhalation
exposure due to radon
52%

Sources of natural background radiation

to be of practical concern, at least compared with other hazards of ordinary life, such as catching the flu. If nothing feasible can be done about them, anyway, they are dismissed as trivial. But doses in the millisievert range deserve attention, and there are some situations where they present significant threats.

For example, frequent long-distance flying on commercial airplanes can bring enough exposure to cosmic rays to require precaution for certain individuals. On the ground, the intensity of cosmic radiation is about a hundred times less than at the four- to seven-mile cruising altitude of passenger jets. Because the shielding ability of the earth's magnetic field is stronger near the equator than at the poles, this radiation increases by a factor of up to three as flight paths move north or south from tropical regions. In addition, due to the varying influence of the sun's magnetic field during the eleven-year cycle of sunspot activity, 40 percent more cosmic radiation reaches earth when sunspots are at minimum than when they are at maximum. Intense solar flares, which shoot charged particles (mostly protons) as well as X-RAYS and gamma rays far into space, can also add to earthly cosmic radiation, but they are rare—the last big one was on February 24, 1956, which produced the most intense cosmic rays ever recorded: 10 mSv per hour at an altitude of 6.2 miles—and of more concern to astronauts. The current solar cycle began in 2008 and is expected to last until around 2020, with a maximum in 2013.

Aircrews and business flyers suffer the highest exposure to cosmic rays—about 60 percent of which is from neutrons and the rest mostly gamma radiation—simply because they spend the most time aloft. OCCUPATIONAL RADIATION surveys have found that European pilots and stewards average about six hundred hours per year, while American crews average as many as nine hundred hours. Business flyers average one hundred hours, but a courier might rack up as many as twelve hundred. At a cruising altitude dose of 5 to 8 µSv per hour in temperate latitudes, everyone on a transatlantic flight between Europe and the United States gets between 30 and 45 µSv (pilots get more, because the passenger cabin provides more shielding than the cockpit). Travel a dozen or so times a year and you collect a quarter of the average annual dose from all natural sources. A nine- to ten-hour haul from New York to Athens approaches 70 µSv. A thirteen-hour flight between New York and Tokyo delivers 75 µSv. Frankfurt to San Francisco can reach 110 µSv. In equatorial zones, the dose rate decreases to 2 to 4 µSv per hour, perhaps making Caribbean vacations even more desirable.

One of many reasons why the Concorde supersonic transport was a rather short-lived extravaganza was that at its cruising altitude of eleven miles, the jet received a cosmic ray dose of 10 to 12 µSv per hour, with upward of 1 millisievert per hour estimated to have occurred if the needle-nosed aircraft had been extant during the 1956 solar flare event. A single one-way trip across the Atlantic would therefore have exceeded the entire average annual dose for people on the ground. During 1990 the average for two thousand flights of the British Concorde was 9 µSv per hour, with a maximum of 44 µSv per hour. French Concordes averaged a bit "hotter," at 11 µSv per hour, because of different routes, but the food was no doubt finer.

Even at subsonic altitudes, the accumulated dose on North Atlantic air routes during a comparatively weak solar event in March 1969 was quite high, at 5 mSv per crossing. There are no government restrictions on air travel based on cosmic ray exposure, but the International Commission on Radiological Protection (ICRP) and the U.S. Federal Aviation Administration (FAA) recommend that pregnant aircrew members limit their exposure to 1.0 mSv for their whole term, with a monthly limit of 0.5 mSv. Other aircrew are limited to a five-year average of 20 mSv per year, with no more than 50 mSv in any single year. Thus, a flight attendant who discovers that she is pregnant at one month and keeps working eighty hours a month on the New York–Athens route would exceed the monthly cap right away and the full-term limit within two months. While it is surely unusual for any women to fly this much, the FAA's advice is worth taking, given the well-established sensitivity of fetuses to radiation and the fact that the mother's body provides no effective shielding. Passengers and crew on board a Chicago-to-Beijing flight during the Halloween 2003 solar storms, which occurred three years after solar maximum and were strong enough to cause a power outage in Sweden, would have received about 12 percent of the ICRP annual limit. No wonder airlines rerouted transpolar flights and flew at lower altitudes during a solar storm in January 2012, which was the strongest since September 2005.

As cosmic rays stream down from outer space, they spawn various RADIONUCLIDES when they collide with atoms in the atmosphere, including TRITIUM (hydrogen-3), BERYLLIUM-7, CARBON-14, and SODIUM-22. These are considered low-LET sources. We wind up ingesting some of them and are thus exposed internally to their radiation. They are negligible except for carbon-14, from which we get an annual dose of about 12 µSv. All in all, the annual worldwide average exposure due to cosmic rays amounts to 0.39 mSv.

Compared with cosmic rays, which comprise only about 16 percent of total annual exposure to natural background radiation, terrestrial sources are somewhat more manageable. Not toiling in an underground mine is a good start.

Greater than half of the 2.4 mSv annual average comes from inhaling RADON-222 and -220 (1.25 mSv), high-LET gases that leak out of the ground and should be abated wherever necessary with proper building construction methods. Radioactive elements remaining from the formation of the earth, such as uranium-235 and -238, thorium-232, RUBIDIUM-87, and potassium-40, have a very long HALF-LIFE and are found in minerals in the earth's crust. Much of the earth's surface where people live consists of soil produced by the weathering of these rocks.

Radon maps are available for many countries, as are distributions of uranium, thorium, and potassium. Exposures to low-LET terrestrial radiation are due mainly to gamma rays emitted from the top ten inches of surface soil and from stony construction materials of buildings—hence the NRC's line about the U.S. Capitol, whose granite blocks are twice as radioactive as typical masonry, due mostly to their potassium content (1,200 BECQUEREL per kilogram of granite versus 500 Bq/kg of masonry). Wood-frame houses are preferable in this regard. Some "hot" locales, such as Kerala and Tamil Nadu in India; Guarapari, Meaípe, and Poços de Caldas in Brazil; and Ramsar, Iran, are notable because of unusually high concentrations of uranium- or thorium-bearing minerals such as monazite. Beach sand in southwestern India contains as much as one-third thorium. Brazil nuts grown in high-gamma-ray soil may be fourteen thousand times more radioactive than most fruits. Population studies of areas with elevated natural background radiation in India and China have not yet found higher risk of cancer, but some show that chromosome aberrations and Down syndrome occur more often than in control groups. In toto, the worldwide annual average exposure is 0.48 mSv from terrestrial radionuclides outside our bodies.

There is still internal exposure to account for. Ludwig von Feuerbach's insight about eating and being (*Der Mensch ist, was er ißt*—"Man is what he eats") came three decades before his countryman, Wilhelm Roentgen, discovered X-rays, so he did not have to meditate upon the transmutation of matter. The annual dose from the fraction of potassium-40 in the body's natural supply of potassium is about 165 µSv for adults and 185 µSv for children. Estimates of exposure from radionuclides in food and drinking water—which come from potassium-40 and the decay of uranium-238 and thorium-232—carry lots of uncertainty, given how much diets vary from person to person and across cultures. LEAD-210 and POLONIUM-210 are the dominant culprits, with the rare PROTACTINIUM-231 and ACTINIUM-227 adding perhaps 1 percent to the total annual dose. Some communities may get more than others. Seafood is relatively high in polonium-210, so would be of more concern in Japan, for example. Indeed, the annual intake of polonium-210 in Japan is ten times greater than in the United States, resulting in human tissue concentrations of 1,200 mBq/kg versus 420 mBq/kg. Fresh fish are generally "hotter" than frozen or canned products, due to radioactive decay. Reindeer and caribou graze on lichens, which bioaccumulate lead- and polonium-210, so lovers of this game are upping their share. So are cigarette smokers, whose lungs contain more lead- and polonium-210 than those of non-smokers; the latter devil has been measured at three times greater in smokers. A five-ounce pack of Brazil nuts brings an exposure of 0.01 mSv from the radium-226 and potassium-40 in the soil where the trees are grown in Central and South America. These are low-LET, but add them all up and the worldwide yearly average is 0.29 mSv from the radionuclides inside us.

How seriously should these admittedly small exposures in everyday life be taken? Because diseases connected with radiation are relatively common and can be caused by other factors, epidemiology has a hard time coming up with crystal-clear evidence of increased incidence tied to low-dose exposure. Many low-dose studies have weak statistical power, because the number of human subjects is small or there are too many complicating factors. The best that science can currently say is that about one individual per one hundred will develop cancer from lifetime (defined as seventy years) exposure just to the low-LET natural radiation sources. Given that background radiation is essentially inescapable, perhaps the most helpful way to regard it is as the irreducible baseline to which all other exposures are added. We do not begin with a clean slate whenever we, say, get a medical X-ray or find ourselves near a nuclear accident.

Because the magnitude of such man-made exposures is often compared to natural background levels by government or industry officials, with the implication that anything close to background can't be all that bad, it is crucial to understand that we never start from zero.

Barium-133, -137, -140

THE ELEMENT BARIUM was named after a Greek word meaning heavy, which is what its natural ores are. The compound barium sulfate ($BaSO_4$) is so dense that it is opaque to X-RAYS and therefore used as a contrast agent for imaging internal organs, especially the gastrointestinal tract. You swallow some of the white powder (or have it pumped in the other end), it coats your guts, and then the doctor studies a sharply defined picture of whatever. For this they are paid hundreds of thousands of dollars a year.

The procedure is not as popular (with physicians) as it used to be, having been supplanted by colonoscopies, ultrasounds, and CT scans. While usually touted as "safe" by the medical profession, a typical barium enema exam entails ten to twenty abdominal X-rays and several minutes of fluoroscopy (continuous real-time imaging, like an X-ray movie), amounting to a radiation dose of at least 6–8 mSv—well above most annual BACKGROUND RADIATION levels around the world. There should be a very good reason for the doctor to ask any patient—especially children and pregnant women—to take such a risk. Everyone should try to keep a record of past history of radiation exposure, so that physicians can better judge the cumulative risk versus the benefit.

Barium has more than thirty RADIOISOTOPES, one of which, barium-130, has such a long HALF-LIFE (7×10^{13} years, longer than the age of the universe) that it is considered primordial and nonradioactive for practical purposes. Barium-133, with a half-life of 10.5 years, is used industrially as a GAMMA RAY source. Barium-137 is nonradioactive but has a briefly unstable precursor that is a DECAY PRODUCT of CESIUM-137, the infamous pollutant from nuclear weapons explosions and reactor accidents. Barium-140, with a half-life of 12.75 days, is a FISSION product that was released in large amounts—some 732 EBq—into the stratosphere by cold war–era weapons tests.

Becquerel

Like the Curie, the becquerel is a radioactivity measurement unit named after a pioneer researcher. The French physicist Henri Becquerel (1852–1908) received the 1903 Nobel Prize in physics along with Pierre and Marie Curie. Though they shared the coveted award, the eternal pecking order of science was evident in the official Nobel citation, which noted that Becquerel had won for his "discovery of spontaneous radioactivity" and the Curies for "their joint researches on the radiation phenomena discovered by Professor Henri Becquerel." Becquerel was the haughty academician, the Curies lowly outsiders. Only Becquerel traveled to Stockholm to receive the award, Madame Curie being too ill from what was not yet understood as the effects of handling RADIUM. When Becquerel later began to pester them for some of the precious element, Pierre wrote to a friend that "we are just fed up with him."

In 1975 the Fifteenth General Conference on Weights and Measures in Paris, which was timed to coincide with the centennial of the 1875 Metric Convention that established an international bureau of standards, "by reason of the pressing requirement, expressed by the International Commission on Radiation Units and Measurements (ICRU), to extend the use of the Système International d'Unités [SI, the metric system] to radiological research and applications, by reason of the need to make as easy as possible the use of the units for nonspecialists, taking into consideration also the grave risks of errors in therapeutic work, adopts the following special name for the SI unit of activity: *becquerel*, symbol Bq."

The benefit of all this to nonspecialists is debatable, but at least the world acquired a radioactivity unit that fit into the metric system. Basically, the old Madame Curie–inspired association with radon was tossed out in favor of a simpler unit defined as the quantity of a radioactive substance in which one nucleus decays per second. Professor Becquerel was not around to bemoan the minusculeness of this unit, as Madame Curie had protested the initial definition of the unit carrying her name. In practice, the becquerel is often found with multiplicative decimal prefixes: *kilo-*, *mega-*, *giga-*, *tera-*, *peta-*, and so on.

Some points of comparison: The radon spring waters at Misara, Japan, contain about 437 becquerels of RADON-222 per liter (Bq/L). The 12.5-kiloton Hiroshima bomb released about 0.1 peta-becquerels (0.1×10^{15}, or 0.1 PBq) of CESIUM-137. About 1,500 PBq of cesium-137 was released into the atmosphere by above-ground testing of nuclear weapons. The Chernobyl reactor disaster released about 89 PBq of cesium-137.

One becquerel equals 2.7×10^{-11} curies, which shows the colossal size of the curie unit that made it rather impractical when it was created in 1910 to honor Pierre and Marie Curie. History has followed suit, with the Curies' popular reputation long ago eclipsing that of their senior professor. *N'importe!* Neither the becquerel nor the curie measures radiation dose, which is handled by the GRAY and SIEVERT.

Beryllium-7, -10

THE ELEMENT BERYLLIUM has a dozen RADIOISO-
TOPES, most with a forgettably short HALF-LIFE, one
with a very long HALF-LIFE (beryllium-10, 1.4 million years),
and beryllium-7, with a human-scale 53 days. None has
much impact on people—the only potential threat is the
beryllium-7 produced in the upper atmosphere when COS-
MIC RAYS collide with oxygen and nitrogen, raining down
eventually on plants that we eat. Beryllium-10 is generated
by cosmic rays, too, but also in nuclear explosions, when
NEUTRONS hit carbon dioxide in the air, which makes it
potentially useful as a tattletale of clandestine bomb tests.
Environmental samples from Hiroshima hold two to three
times more beryllium-10 than natural cosmic ray deposits
would account for.

Beta particles

IKE BULLETS from a hair-trigger-loaded gun, beta particles are electrons given off by the unstable nuclei of certain radioactive elements. The energy they carry can be great enough to bust molecules apart, making them harmful to living cells.

If Henri BECQUEREL, who discovered radioactivity in 1896, and Ernest Rutherford, who named the beta particle in 1899, had known that this form of radiation was composed of electrons (which had been named in 1894 and identified in 1897), perhaps there would never have been a separate term for it. But when Rutherford found that Becquerel's mysterious "rays" emitted by uranium salts were made of particles, he used the Greek letters *alpha* and *beta* in order to differentiate their ability to penetrate matter. ALPHA PARTICLES are slower and 7,500 times heavier than beta particles—a sheet of newspaper will halt them. But it takes several millimeters of aluminum foil and more than three feet of concrete to stop beta particles. After Marie and Pierre Curie found the electrical charge of beta rays to be negative in 1900, Walter Kaufmann of the University of Göttingen proved them to be electrons in 1902.

The electrons in beta radiation, which travel near the speed of light, are of much higher energy than those that surround an atom's nucleus. Even so, they may not rid an unstable nucleus of all the energy it needs to shed in order to become stable. GAMMA RAYS, which are more penetrating than either alpha or beta particles, therefore often accompany beta decay to help bleed off more energy. Thick layers of lead are needed to block gamma radiation. More exotic subatomic particles, such as neutrinos, are also part of beta emission, but they are not relevant here for understanding the health effects of radiation.

Beta particles can travel several feet in air before they lose most of their initial energy and become just like any free electron that eventually neutralizes its negative charge by joining up with something positive. Because they can travel farther into living tissues than alpha particles, they are considered more dangerous as an external source of exposure, easily capable of burning the skin like a flame (known as a "beta burn"). Marshall Islanders and the crew of a Japanese fishing boat named *Fukuryu Maru* (Lucky Dragon) suffered extensive beta burns from FALLOUT after an American H-bomb test at Bikini Atoll in 1954. Beta particles are extremely dangerous if somehow ingested, of course, though not as terrible as the larger and more strongly charged alpha particles in this regard.

Beta emitters often mentioned in the news and

encountered—rarely, one hopes—in the environment include CESIUM-137, COBALT-60, IODINE-129 and -131, and STRONTIUM-90, each of which also emits gamma radiation. POTASSSIUM-40, which is ubiquitous in plants and animals, is an important natural beta emitter.

Beta particles can sometimes generate X-RAYS—specified by the German word *bremsstrahlung*, or braking radiation—as they slow down and career past heavy atomic nuclei. High-energy beta emitters such as PHOSPHORUS-22 are therefore shielded with relatively low density materials such as Plexiglas or water rather than lead.

Bismuth-205, -210, -212, -213, -214

BISMUTH BOASTS ONE of nature's zombie RADIOISO-
TOPES, one whose heart beats so slowly that it might as
well be dead. With a HALF-LIFE of 1.9×10^{19} years—a billion
times longer than the age of the universe—bismuth-209
emits ALPHA PARTICLES and decays into THALLIUM-205.

There are dozens of other, much livelier bismuth radioisotopes. Bismuth-210, at five days, is a DECAY PRODUCT of
URANIUM-238 and therefore of concern in the operation
of nuclear reactors. It was known historically as "radium E"
because of its position in the stages of RADIUM disintegration, as that element transmutes into the POLONIUM-210
discovered by Marie and Pierre Curie in 1898 ("radium C"
was also a bismuth isotope, bismuth-214, with a half-life of
just twenty minutes). Bismuth-212, a dangerous alpha emitter with a half-life of one hour, is a decay product of
THORIUM-232, which means that it is found around thorium's dangerous daughter, RADON-220, aka thoron. Bismuth-
214 is itself a daughter of RADON-222 and accounts for some
of that ubiquitous gas's health hazard due to its ability to
lodge as a solid particle in lung tissue.

Only one member of the bismuth gang has beneficial traits. Bismuth-213 can be used to treat leukemia and other cancers by attaching it to a monoclonal antibody, which is injected into patients and dispatched to bombard cancer cells with high-energy alpha particles. This experimental procedure is known as targeted alpha therapy, or TAT.

Boron, cadmium

IN ORDER TO TAME the FISSION reactions in a nuclear reactor—that is, practically speaking, to turn the power up and down or on and off—the number of NEUTRONS bombarding the core's fuel must be controlled. The more neutrons there are, the more fission, the more heat, the more electricity generated. The elements boron and cadmium are excellent neutron absorbers, so they are mixed into durable alloys to make rods that can be raised or lowered between other rods containing ENRICHED URANIUM fuel. When the rods are fully inserted, the reactor stops producing energy, because there are not enough neutrons to split the uranium nuclei in a self-sustaining chain reaction—the reactor is said to be below *criticality*. As they are retracted, it starts up and heads toward full, steady power. An emergency shutdown, which relies on several redundant mechanisms to drop or thrust the control rods automatically into the core in a matter of seconds, is called a *scram*, a term that dates back to the first Manhattan Project reactor, at the University of Chicago in 1942.

Natural boron is made of two nonradioactive isotopes,

boron-10 (20 percent) and boron-11 (80 percent). Because boron-10 is the better neutron sponge, boron control rods contain material that has been artificially enriched to be almost pure boron-10. Boron is also used in shields that keep components near the core from becoming radioactive from the neutrons streaming out of it. To prevent spent fuel rods from fissioning, they are placed on underwater racks with boron carbide plates that act as neutron absorbers, and soluble boron is mixed in the pool. Similar boron solutions, called "neutron poisons," are stored in tanks close to the reactor core in case of a total failure of the control rod mechanisms. During the 1986 Chernobyl reactor disaster, sodium borate (borax) and sand were dumped by helicopters onto the openly burning core. A total of about forty tons of boron compounds were dumped from above before the mission was halted because of high radiation levels and inaccuracy. At the tsunami-damaged Fukushima Daiichi plant in 2011, seawater mixed with borax was pumped into the damaged reactor chambers in a desperate attempt to cool them down. In November 2011, when the discovery of XENON-133 and -135 in gas samples from reactor 2 at the plant raised fears that melted fuel in the wrecked core was sustaining fission again (a phenomenon called recriticality), workers injected water mixed with boric acid into the reactor.

Quaffing various boron compounds as a way to prevent radiation exposure is quack medicine. In fact, boron is considered to be a chemical hazard in drinking water.

Bromine-77

RADIOACTIVE SUBSTANCES are never bromides. The element bromine has a couple dozen radioisotopes, many of which are worrisome as FISSION products, but all have a mercifully short HALF-LIFE. At fifty-seven hours, bromine-77 lasts the longest. None was ever present in that classic hangover cure from Baltimore, Bromo-Seltzer, whose sodium bromide content was in any case outlawed by the Food and Drug Administration in 1975.

Calcium-45, -47

LIKE CARBON, CALCIUM IS NECESSARY for life. Because of calcium's many physiological roles as a vital constituent of bones and a controller of muscles and nerves, the man-made RADIOISOTOPES calcium-45 and -47 have found applications as tracers injected into the bloodstream for the study of metabolic pathways and the medical diagnosis of calcium disorders. Calcium-47 is preferred, because of its shorter HALF-LIFE (4.5 versus 163 days). Both emit BETA PARTICLES. Patients who receive these radioactive beacons might truly be described as in the limelight.

Calcium-45 has also been used to track what happens to calcium in soils treated with various fertilizers.

Californium-252

CALIFORNIUM IS A MAN-MADE ELEMENT born in a cyclotron in 1950 under the parentage of RADIOISO-TOPE progenitor Glenn Seaborg and colleagues at the University of California, Berkeley. Naming their baby after the state and the university was a departure from the custom of relating a brand-new element linguistically to the one right above it in the periodic table—in this case, dysprosium (from the Greek word for "difficult to get at"), whose French discoverer needed more than thirty tries to isolate it in 1886. The Berkeley wags justified their patriotic coinage by recalling that prospectors for gold had never found it easy to get to California. As with every artificial element, all of californium's isotopes are unstable, like much of life in the Golden State.

As a by-product of PLUTONIUM production, californium was saved from being a major environmental pollutant because it was never spawned in sizeable quantities. Above-ground testing of nuclear weapons during the cold war era deposited small amounts in FALLOUT. While nominally an emitter of ALPHA PARTICLES, californium-252 is also an intense source of NEUTRONS, due to its high rate of spontaneous FISSION—that is, sometimes (about three out of every hundred decays) it just splits apart and spews neutrons. With a HALF-LIFE of 2.64 years, this makes it both dangerous and useful for industrial applications, such as nuclear reactor startup, cancer therapy, oil well analysis, and—*eureka!*—gold prospecting, wherein stable natural gold (Au-197) is transmuted into detectably radioactive gold (Au-198) by the neutrons from a portable californium-252 source. Better than panning by hand, pardner.

Carbon-11, -14

THE ELEMENT CARBON IS EVERYWHERE. One of its RADIOISOTOPES, carbon-14, is found in every living thing. There are about a dozen other radioactive carbon isotopes, but carbon-14's HALF-LIFE of 5,700 years is by far the longest. Carbon-14 is created constantly, at about 1 peta-BECQUEREL per year when NEUTRONS from COSMIC RAYS collide with nitrogen atoms in the upper atmosphere, where it then combines with oxygen to make carbon dioxide. Plants in turn absorb the $^{14}CO_2$, animals eat the plants, we eat those animals as well as the plants, and thus carbon-14 accumulates in the food chain to contribute to natural BACKGROUND RADIATION exposure. The average annual global EFFECTIVE DOSE from cosmogenic carbon-14 is about 12 microsieverts.

Carbon-14, a BETA PARTICLE emitter with no attendant GAMMA RAYS, was also dumped into the atmosphere by above-ground nuclear weapons tests during the cold war era—a total of about 220 PBq, mostly in the Northern Hemisphere. In a fashion analogous to cosmic rays, fusion (H-bomb) explosions released neutrons that reacted with nitrogen in the air. This was by far the largest contributor of any RADIONUCLIDE to the worldwide collective dose from bomb FALLOUT. The average annual effective dose from this activity peaked in 1964, at around 7.7 µSv, and is about four times less than that now. Today the yearly dose from lingering weapons fallout is about equally divided between external irradiation and ingested sources, with ingested carbon-14 accounting for about a third of the total, which is more than the internal dose from all other fallout radionuclides (such as STRONTIUM-90 and CESIUM-137).

Like many of the radioisotopes discovered during World War II, carbon-14 later found nonmilitary applications. So-called radiocarbon dating—pioneered by the American physical chemist Willard Frank Libby (1908–1980), who went from developing ENRICHED URANIUM for the Hiroshima bomb to winning the 1960 Nobel Prize in chemistry for revolutionizing the dating of fossils and artifacts containing organic matter—relies on measuring the decreasing ratio over time of carbon-14 to stable carbon-12 in a dead organism.

Carbon-11 is used as a tracer in medical PET (positron emission tomography) scanning, which produces 3-D images of the body. Its twenty-minute half-life minimizes the radiation dose to the patient.

Cell phones

SCIENTIFIC JARGON CAN BE USEFUL for scientists but bewildering for laymen. More often than everyday language, it has strict definitions. The word *radiation*, for example, is applied technically to a spectrum of energy that includes visible light, radio and television signals, radar, and X-RAYS. In the popular mind, the Atomic Age has connected radiation forever with gruesome injuries from FALLOUT and the risky benefits of cancer therapy. That a new consumer gadget enjoyed by an estimated five billion people worldwide can be said to emit radiation is both a technical fact and a confusing issue.

Cell phones are low-power (less than 2 watts) radio transmitters that operate at what engineers call microwave frequencies, sending out electromagnetic energy that travels in waves between 450 million and 2,700 million cycles per second (or 450–2700 megahertz, abbreviated MHz). When they are held against the ear to make a voice call, this energy is absorbed by the brain and skull. Unlike X-rays, however, whose waves move at enormously higher frequencies— nominally between 30 petahertz and 30 exahertz—

microwaves are not strong enough to break molecules when they hit living tissue. Scientists say that microwaves are *non-ionizing* radiation. They might be able to heat substances up—this is how microwave ovens work—but they cannot destroy cells, damage DNA, or tear anything apart on the level of chemical bonds.

During the past two decades, as cell phones have grown from a communications novelty to a personal necessity, numerous studies have tried to find out if they pose any threat to health. So far, the answer is "no, but let's keep looking." Cell phones do affect brain activity when held against your head, but the ramifications are unclear. The biggest study so far, called Interphone, examined data from thirteen countries and found no consistent link between using mobile phones for more than ten years and the presence of head or neck tumors among adults. A study led by the Institute of Cancer Epidemiology in Denmark, which tracked 358,403 people with cell phones for eighteen years, found no significant difference in rates for brain or central nervous system cancers among users and nonusers. Such large projects are notoriously difficult to keep free of biases and reporting errors, and can so far assess only cancers that would develop over a rather short period of time, but even animal and in vitro studies have not reliably shown increased risk of cancer from long-term use of cell phones.

Nonetheless, there are enough worrisome fragments—for example, a 40 percent increased risk for glioma among the heaviest users (average thirty minutes a day) in the

Interphone study—to lead the World Health Organization to play it safe by classifying the electromagnetic fields produced by mobile phones as "possibly carcinogenic to humans." WHO is scheduled to complete a formal review of all existing studies in 2012. The British Department of Health advises that children under sixteen should use mobile phones "for essential purposes only" and keep calls short (good luck, mates!).

It is possible to estimate how much electromagnetic radiation your body receives from various cell phone models. A standardized measurement obtained under laboratory test conditions—the specific energy absorption rate, or SAR—is expressed in watts per kilogram of tissue. In the United States, the Federal Communications Commission (FCC) requires cell phones to be at or below 1.6 W/kg, a conservative threshold meant to prevent heating effects. The iPhone 4S, for example, has a SAR rating of 1.18 to the head in the CDMA 1900 MHz band. Check the fine print of your phone's user manual.

Meanwhile, talk less, text more. Or read a book.

Cerium-141, -144

CERIUM-141 AND -144 ARE FISSION products with half-lives of 32.5 and 285 days, respectively. In the early 1960s, small amounts of cerium-144 from above-ground nuclear weapons test FALLOUT were found in samples of food and animal bone from Ibaraki, Japan. The highest level of radioactivity was detected in clams. American wheat and flour were also discovered to be contaminated. In the first days after the 1986 Chernobyl reactor disaster, human respiratory problems were linked to aerosol forms of cerium-144 released into the atmosphere as fuel particles. Large particles, varying in size up to tens of micrometers, containing cerium-141 and -144 and other radioactive debris were later collected in Bulgaria, Finland, Germany, and Hungary. Cerium-144 was found along with IODINE-131 in highly radioactive water (400 mSv per hour) that inflicted BETA PARTICLE burns on emergency workers at the damaged Fukushima Daiichi reactor plant in March 2011.

Cesium-134, -135, -136, -137

THE ELEMENT CESIUM was known long before discoveries about radioactivity ushered in the modern age
of physics, but several of its RADIOISOTOPES haunt the field
like vile children. In 1860 the German chemist Robert
Bunsen (1811–1899)—inventor of the eponymous gas-flame
"burner" familiar to chem lab students everywhere—
distilled forty tons of Bad Dürkheim mineral water to obtain
less than half an ounce of a new element he named "cae-
sium," from the Latin word for bluish-gray color. Cesium's
radioactive isotopes, none of which occurs in nature, were
not cataloged until eight decades later, during the era when
American nuclear chemist Glenn Seaborg turned out dozens of
artificial RADIONUCLIDES under Manhattan Project auspices.

The most infamous brat of the forty-member cesium
gang is surely cesium-137, a dangerous FISSION fragment
that has been jettisoned into the biosphere by nuclear reactor
accidents and weapons tests. Much smaller quantities routinely escape in "planned releases" from power plants during
maintenance or refueling. It emits BETA PARTICLES and decays into a barium radionuclide that gives off strong GAMMA

RAYS, thus creating both an internal and external hazard that can cause myriad kinds of cancer. With a HALF-LIFE of thirty years, it takes more than a century to decay past the worry point. It is a long-term soil contaminant, carried by tainted food into the human body, which welcomes it as though it were natural potassium—a nasty ruse.

In August 2011 the Japanese government estimated that the amount of cesium-137 released by the stricken Fukushima Number 1 plant—1.5×10^{16} BECQUERELS—was equivalent to 168 Hiroshima bombs. Independent analysis by scientists put the total as much as five time higher, with about a fifth of the FALLOUT deposited over Japanese land areas and most of the rest over the North Pacific ocean. The total amount of cesium-137 released by weapons testing from 1945 through 1980 was about 9.6×10^{17} Bq, mostly in the Northern Hemisphere. The 1986 Chernobyl reactor disaster dumped about 8.5×10^{16} Bq, with the heaviest impact on the European part of the former USSR, where about 3 percent of the land—home to more than five million people—was contaminated with cesium-137 at greater than 37 kBq per square meter (areas with greater than 555 kBq/m^2 were designated as strict-control zones, receiving regular monitoring and preventive measures). Trace amounts can now be found all over the world.

The other cesium radioisotopes are less devilish. Cesium-134's half-life is just two years, so it is threatening only in the shorter aftermath of a reactor accident. Cesium-135 stays around for an astronomically long time, with a half-life of

2.3 million years, but its beta particles are low-energy and there is no gamma radiation. Cesium-136 has a half-life of just thirteen days, but has been detected in samples of human breast milk from northern Sweden at concentrations as high as 178 picoCURIES per kilogram.

In December 2011, cesium-134 and -137 were found in powdered milk manufactured by Meiji, Japan's largest supplier of infant formula, at concentrations as high as 15.2 Bq/kg and 16.5 Bq/kg, respectively. The regulatory limit is 200 Bq/kg, but Meiji nonetheless recalled some 400,000 cans of formula that had been made at a factory in Saitama prefecture, about 200 kilometers southwest of Fukushima, during the week after the March 2011 reactor meltdowns. According to Japan's National Institute of Radiological Science, daily consumption of the product would have produced an exposure of about 0.07 microSIEVERTS in a nine-month-old baby. Shipments of beef and rice from the Fukushima region had already been banned due to radio-cesium contamination. Fallout from the damaged reactors has been found in all of Japan's prefectures, with highest cesium concentrations in an oblong area reaching about fifty kilometers northwest of the plant.

There are some relatively benign uses of cesium radioisotopes, which might bring occupational exposure for workers but are otherwise not worrisome. Cesium-137 is a gamma ray source for the sterilization of wheat, flour, potatoes, surgical equipment, and other medical supplies. It is also used in industrial radiography (which uses gamma rays to obtain images similar to X-RAY photos): for inspecting oil pipelines and for seeing through transport containers at border crossings. Because of its commercial availability, it is regarded as a potential terrorist agent.

Chromium-51

CHROMIUM-51 IS A MAN-MADE RADIOISOTOPE with a HALF-LIFE of twenty-eight days used as a tracer in medical studies of red blood cells and gastrointestinal functions. Occupational exposure to chromium-51 among nuclear workers has been associated with increased risk for prostate cancer.

Cobalt-57, -58, -60

YET MORE OF AMERICAN NUCLEAR chemist Glenn Sea-borg's prodigious output of man-made RADIOISOTOPES from the late-1930s and 1940s, a prolific period fueled by the Manhattan Project that led him to the 1951 Nobel Prize in chemistry. Cobalt has been used for thousands of years to color blue glass and ceramics but was not identified as an element until the mid-eighteenth century. Medieval German miners believed that hobgoblins they called kobolds lurked in underground shafts, sometimes tricking them to dig for noxious ores. A bit of scientific reality lay behind their animism: rocks containing cobalt also carry arsenic, which is released during smelting.

This etymology suits cobalt's radioactive isotopes better than any old-time *Bergmann* (miner) could have imagined. More than twenty of them exist as artificial creations, with cobalt-57 and -60 of commercial industrial interest because of their longer half-lives (272 days and 5.27 years, respectively). Both emit BETA PARTICLES and GAMMA RAYS. Cobalt-60 is best known as a source of energetic gamma radiation for radiography, bacterial sterilization of food, and radiotherapy. It is also present (along with cobalt-58) in spent nuclear fuel and other radioactive waste from nuclear power plants, and in FALLOUT from the bygone era of atmospheric nuclear weapons tests.

Given the ubiquity of radiotherapy machines in hospitals and clinics around the world, it is not surprising that they have been at the center of infamous accidents. In 1983, six thousand pellets of cobalt-60 from a scrapped radiotherapy unit in Juárez, Mexico, found their way from a junk hauler's truck into hundreds of tons of steel products manufactured by local foundries and shipped throughout the American Southwest—everything from reinforcement bar to patio tabletops. The blunder was discovered only when a truck delivering rebar to the Los Alamos National Laboratory in New Mexico took a wrong turn and passed by a roadside radiation monitor. What goes around comes around—in this case, about 400 CURIES. Much, but not all, of the hot steel was recovered. In 1993, two disused cobalt-60 radiotherapy sources in Ankara, Turkey, that were repackaged for export to the United States somehow came to be stored in an empty Istanbul warehouse. Five years later, two scrap haulers dismantled the containers at home, where they quickly got sick with nausea and vomiting. A total of eighteen people were soon admitted to hospital, where ten suffered from ACUTE RADIATION SYNDROME (ARS) and five were treated for forty-five days. In 2000, a junkyard worker in Samut

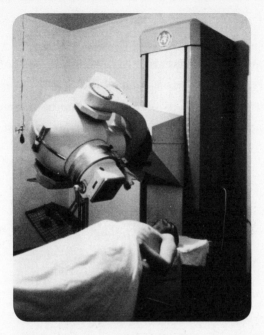

A cancer patient exposed to cobalt-60

Prakan, Thailand, cut open a cobalt-60 source with a welding torch, which led to the exposure of 1,870 people who lived near the dump, ten of whom developed symptoms of ARS that killed three of them within two months.

In January 2012, Bed, Bath & Beyond stores across the United States recalled cobalt-60–tainted metal tissue holders imported from India, after packages containing them triggered radiation alarms at truck scales in California. A spokesman for the Nuclear Regulatory Commission said that someone spending thirty minutes a day for a year near one of the boxes on a bathroom vanity would receive the equivalent of a couple of chest X-rays. For free!

Those kobolds never rest. Excess heart disease mortality has been observed among women with breast cancer who were irradiated with cobalt-60. It has also been associated with increased risk of prostate cancer among nuclear workers.

Containment

NOT TO BE CONFUSED with the name for American foreign policy toward the Soviet Union during the cold war, though the concept is similar, containment refers to structures built to keep radioactivity from escaping into the open environment around nuclear reactors. Engineers know that there is no such thing as a perfect seal, so no containment is 100 percent tight. History has shown that containment structures work as designed until something happens that the designers did not foresee, at which point the most salient difference between nuclear and conventional power plants becomes obvious to everyone: a breach can make wide areas uninhabitable for many years. The containment around American reactors was never intended to hold melted fuel, because safety regulators forty years ago believed meltdowns were impossible.

Constructed of concrete and steel, containment buildings of the most common reactor type, the pressurized water reactor (PWR), can be recognized by their large aboveground cylindrical or spherical profiles, which offer the best geometry for preventing high-pressure radioactive gases from leaking outside in an accident. A boiling water reactor (BWR) surrounds the core itself with walls made of the same materials, but all that is usually visible from outside is a boxy "secondary containment" building, which is essentially a lightweight roof over the reactor system. The Three Mile Island reactor in Pennsylvania, where radioactive gases were vented outside during a meltdown in 1979 when pressure soared to critical levels, was a PWR. The Chernobyl reactor, which—incredible though it may now seem—had no hard containment building, and the Fukushima reactors, where two secondary containment buildings were blown apart by violent hydrogen explosions, were both American-designed BWRs. Fuel in the Fukushima reactors not only melted down, but melted through their inner steel containment vessels and onto the floor of their concrete outer containment.

Six months after the Fukushima disaster, the U.S. Nuclear Regulatory Commission announced that it would "re-examine the level of conservatism" in the seismic design of American commercial nuclear reactors. That is, it would see if the reactors had been built strong enough originally. A 5.8 magnitude earthquake in Virginia on August 23, 2011, in a region where such strong temblors are rare, pushed nearby reactors beyond their design limits. Dominion Resources, the utility company that operates the 1970s-vintage North Anna plant fifty-two miles northwest of Richmond, Virginia—which automatically shut down when the quake apparently caused a problem inside the reactor cores—

Typical Boiling-Water Reactor

How Nuclear Reactors Work

In a typical design concept of a commercial BWR, the following process occurs:

1. *The core inside the reactor vessel creates heat.*
2. *A steam-water mixture is produced when very pure water (reactor coolant) moves upward through the core, absorbing heat.*
3. *The steam-water mixture leaves the top of the core and enters the two stages of moisture separation where water droplets are removed before the steam is allowed to enter the steamline.*
4. *The steamline directs the steam to the main turbine, causing it to turn the turbine generator, which produces electricity.*

The unused steam is exhausted to the condenser, where it is condensed into water. The resulting water is pumped out of the condenser with a series of pumps, reheated, and pumped back to the reactor vessel. The reactor's core contains fuel assemblies that are cooled by water circulated using electrically powered pumps. These pumps and other operating systems in the plant receive their power from the electrical grid. If offsite power is lost, emergency cooling water is supplied by other pumps, which can be powered by onsite diesel generators. Other safety systems, such as the containment cooling system, also need electric power. BWRs contain between 370–800 fuel assemblies.

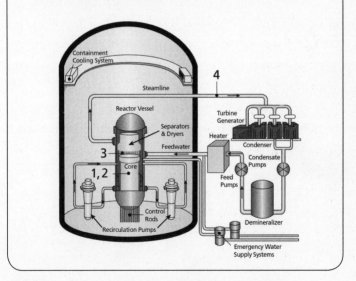

Containment, at least in theory, around a boiling water reactor

reported to the NRC that "cumulative absolute velocity," a measure of shaking, exceeded the ground motion for which the plant was designed. The NRC's own measurements estimated the shaking at 26 percent of the force of gravity, while the containment buildings were designed to withstand just 18 percent. North Anna was the first reactor to shut down during an earthquake in the fifty-three-year history of American commercial nuclear power.

Cosmic rays

C OSMIC RAYS WERE NAMED in 1925 by the American physicist Robert Millikan (1868–1953), but someone else had already discovered them in 1912. After Father Theodor Wulf (1868–1946), a German physicist and Jesuit, carried a primitive radiation detector to the top of the Eiffel Tower in 1910 and recorded higher readings up there than on the ground (everyone had expected *lower* numbers), an Austrian physicist named Victor Hess (1883–1964) took measurements from a balloon in 1912. The radiation intensity three miles high was twice that on terra firma. Hess concluded that the radiation was coming from outer space, but used the apparently un-catchy term *ultra-radiation* for it. Nonetheless, he shared the 1936 Nobel Prize in physics with Carl Anderson (1905–1991), who discovered the positron while studying cosmic rays with Millikan (who scored his own Nobel in 1923 for figuring out the charge of the electron). Hess's otherwise illustrious career was darkened by his position as director of research from 1921 to 1923 at the infamous U.S. Radium Corporation in New Jersey, which at that time was poisoning its women watch dial painters with radium-based paint. Small world.

All cosmic rays come from Out There, with those of highest energy coming from so far beyond our galaxy that they started before the earth existed. These galactic cosmic rays are about 86 percent protons, 12 percent ALPHA PARTI-CLES, and the rest carbon, oxygen, and rarer nuclei, plus some electrons and positrons. Supernovas probably created them, but the rare ultra-high-energy ones are the stuff of theory. At a lower energy level, protons from solar flares can produce significant radiation exposures at high altitudes, but few of them reach the ground. The earth's magnetic field and atmosphere shield life from most of this radiation, thus making life possible.

At sea level, cosmic rays account for about 10 percent of natural BACKGROUND RADIATION, 10 percent higher near the poles than at the equator. The dominant charged parti-cle is actually the muon—an unstable fundamental particle (also discovered by Carl Anderson) created by atomic colli-sions as cosmic rays stream down through the atmosphere—which has the same negative charge as an electron but is two hundred times heavier. Muons account for around 80 per-cent of the average annual effective dose of about 0.27 mSv from cosmic rays.

About a dozen RADIONUCLIDES of interest are produced when cosmic rays collide with elements in the atmosphere. TRITIUM and CARBON-14 are constantly churned out by the interaction with oxygen and nitrogen. Most of the tritium winds up in seawater and causes little human exposure. On the other hand, carbon-14 can be found in every compound in the human body. The average annual dose from these radioisotopes is about 0.01µSv and 12µSv, respectively. BERYLLIUM-7 comes down in rainwater and contributes about 0.03µSv via the consumption of fresh vegetables. Finally, SODIUM-22 brings another 0.15 µSv per year. There is truly nothing to be done about these exposures, and they should be on no one's worry list.

Critical mass

A TERM FROM NUCLEAR WEAPONS history that, like FALLOUT and *megaton*, has leaked into the lingua franca. A critical mass of URANIUM-235 or -233, or PLUTONIUM-239 (substances that are stable enough to be stored practically, yet can be readily split apart to release explosive energy) is the minimum quantity that will sustain a cascade of FISSION reactions, whereby free neutrons break the material into fragments of other elements plus new free neutrons, along with a tremendous release of pure energy. Knowing what this quantity is enables the design of a bomb that will blow up only when you want it to, because anything less than the critical mass will not work. At the beginning of World War II, physicists calculated it and realized that a terrible new weapon would be possible.

The Hiroshima bomb carried about 60 kilograms of uranium-235 (80 percent ENRICHED URANIUM), though an atomic explosion could have been created with just 15 kilograms. Two stable "subcritical" pieces were fired rapidly together in a cannon-like tube to create an explosive critical mass. The Nagasaki bomb carried 6.2 kilograms of plutonium in a low-density sphere—too low to sustain fission—that was made denser by being compressed by shock waves from conventional explosives in order to fission about 1 kilogram of it. Most modern nuclear weapons use plutonium.

Some of the best minds in twentieth-century science spent the prime years of their lives in the Manhattan Project during World War II, building history's most destructive weapon. "Now I am become Death, the destroyer of worlds," Robert Oppenheimer, their leader, recalled remembering from the *Bhagavad Gita* upon witnessing the first test. Critical mass was their trump card.

In a nuclear reactor, the same fission reactions with uranium or plutonium are used to create heat for making steam that will then turn electric turbine generators. In this case,

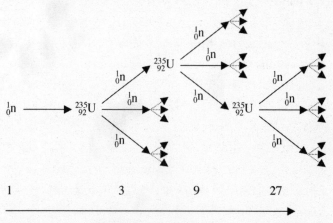

Number of neutrons

Neutron avalanche during fission of a critical mass of uranium-235

the fuel is not enriched enough to obtain bomb-grade concentrations of the necessary radioisotopes that would cascade
at an explosive rate. A nuclear reactor can therefore never be
a bomb, though it creates the same radioactive fragments
that are dangerous if released into the environment. The
reactor core is said to be "critical" when controlled fission is
maintained in normal operation.

Curie

EARLY RESEARCHERS WERE STUNNED by the fact that radioactive material did not stay the same from moment to moment like ordinary stuff. It gave off particles and energy and changed in fundamental ways, sometimes into entirely different elements. It was *active*, hence the term itself: radio-*activity*. Needing some standardized way to measure this phenomenon, the 1910 International Congress of Radiology and Electricity in Brussels convened a Radium Standards Committee chaired by the great physicist Ernest Rutherford, which created a unit named the curie after pioneer researchers Marie and Pierre Curie. Science was still one of the most sexist professions on earth, so some participants believed they were naming it after just Pierre, others after Marie, but cagey Rutherford seems to have kept this contentious social issue ambiguous.

The committee first decided that one curie would equal the quantity or mass of RADON that exists when the decay products of 10^{-8} grams of RADIUM reach a steady state. But none other than Marie Sklodowska Curie herself protested that this was far too minuscule to honor her husband's name, insisting that a whole gram of radium should be the basis instead. That a gram of pure radium in those days was a fantastic quantity then worth about $70,000—typical laboratory amounts weighed a few milligrams, with a ton of pitchblende ore (UO_2) necessary to obtain just one-tenth of a gram of radium chloride—meant that such a unit would be totally impractical, but Madame Curie was not to be denied.

Today the curie (symbol Ci) is defined not in terms of radon but as the quantity of any radioactive substance undergoing a certain number of decays, or nuclear disintegrations, per second: 3.7×10^{10}, or 37 billion. This is roughly the *activity* of one gram of RADIUM-226, thus maintaining the historical link to Marie, or Pierre, or both of them. Thus, one curie of radium is about one gram of radium.

In the confusing constellation of radiation measurement units, it helps to remember that the curie is not a measure of exposure or dose (see GRAY, SIEVERT), nor of the amount of any certain type of radiation emitted by a substance, but rather, a raw physical gauge of radioactivity itself at the atomic level. For example, a typical medical radiotherapy machine for killing cancerous cells with GAMMA RAYS employs a source consisting of COBALT-60 pellets at greater than 300 Ci per gram. The device may be rated at a maximum 8,000 Ci.

One curie equals 37 billion BECQUERELS (the other common unit of radioactivity). Perhaps this would please

Marie Curie, her youthful beauty already marred by fatigue from radiation exposure

Madame Curie, who never much liked the imperious Professor Henri Becquerel, a member of the same French Academy of Sciences that would not admit a woman to membership until 1962. In 1995, when her ashes and Pierre's were placed in the Pantheon, France's national mausoleum (in separate tombs), Marie Curie became the first woman so honored among the nation's "great men." Better late than never.

She had already paid dearly for any such honor. As early as January 1903 her husband wrote to a friend that she was "always very tired, without being really ill." A year later he wrote that "she had a miscarriage after five months of pregnancy and since then she is weary and tired." Without doubt, these were effects of her handling radium and breathing radon gas.

Curium-242, -243, -244

THE MAN-MADE ELEMENT CURIUM was created by Mr. RADIOISOTOPE, Glenn Seaborg, and his team of Manhattan Project researchers at the University of California, Berkeley, in 1944. They were hell-bent on producing PLUTONIUM for an atomic bomb and therefore kept secret the by-product curium, which they named after Pierre and Marie Curie (who was a lifelong pacifist), until after Nagasaki and the end of World War II. It is reasonable to assume that she would not have approved.

All of curium's twenty-some isotopes are radioactive, but only three have half-lives that make them both troublesome and potentially useful: curium-242 (163 days), curium-243 (29 years), and curium-244 (18 years). Each is a potent emitter of ALPHA PARTICLES and each transmutes into plutonium. They are rare, expensive, and highly radio-toxic, which means they have few applications beyond research. Curium-242 was used as a source of alpha radiation in early lunar missions that determined what the moon's soil was made of by analyzing the scatter of the emitted particles. Mars rovers have also carried this technology.

FALLOUT from atmospheric nuclear weapons tests and reactor accidents has deposited curium in the environment. Small quantities of curium-242, -243, and -244 were detected in playground soil near the Fukushima Daiichi plant in Japan after the March 2011 tsunami disaster. Its biological HALF-LIFE is twenty years—a long, long time for children.

Decay product

THE FIRST DISCOVERIES OF RADIOACTIVITY in the 1890s and 1900s caused a sensation far beyond rarefied academic laboratories because the property seemed magical, harkening back to the ancient alchemist's power to change one kind of matter into another. Marie Curie's amazing radium even glowed in the dark, an irresistible draw for quack medical treatments and luminous consumer products from watches to women's dresses. It took years for physicists to figure out what was actually happening, let alone for everyone to accept that radium could be dangerous, but the transformation of radioactive elements into a chain of completely different elements, called decay products, remains one of the wonders of nature.

For example, the URANIUM-238 that makes up most of the natural uranium left from the earth's creation decays into nine different elements in the following order: thorium, protactinium, radium, radon, polonium, lead, astatine, bismuth, and thallium. The final result is a bit of stable lead—not the gold that alchemists sought, but a pretty good trick. Sometimes the decay products are much more

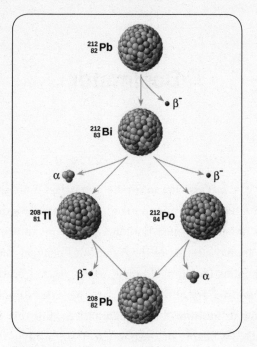

Thorium decay chain from lead-212 to lead-208

radioactive than the original element. Depending on the HALF-LIFE of each decay product, the process may be nearly instantaneous or take billions of years. Thus, the radioactivity of a chunk of pitchblende ore containing uranium-238 changes over time, rising with how much highly radioactive radium is present.

The male-dominated world of science used to refer to decay products as the "daughters" or "progeny" of a radioactive parent, but this is no longer fashionable.

Dosimeter

"MY RULE IS not to expose in ten days more than the number of minutes required to produce a dermatitis," a physician named Ennion G. Williams informed the 1905 annual meeting of the American Roentgen Ray Society at Johns Hopkins Hospital, which was celebrating the tenth anniversary of the discovery of X-RAYS. No one knew yet how to measure X-rays quantitatively, but this did not stop Dr. Williams or anyone else from aiming them at patients, despite widespread evidence that they could be very harmful. By World War I, various chemical reactions caused by X-rays that produced color changes were being used to calibrate crude dosimeters—which resembled watercolor paint sets—in macabre units defined as "one-third of the radiation necessary to set up the first signs of reaction in the healthy skin of the face." Perhaps it was better than nothing, but not much.

Today, workers whose jobs may entail exposure to radiation (including doctors) routinely wear portable meters to record the doses they receive. These dosimeters do not protect against radiation but rather, simply measure cumulative

May I borrow your dosimeter, sir?

exposure after a period of time. They are usually in the form of detector badges that clip on to clothes and measure X-rays, GAMMA RAYS, and higher-energy BETA PARTICLES. Astronauts wear several.

There are three common types of dosimeters, known as film badge, quartz fiber, and thermoluminescent, which operate on different physical principles. They are too expensive to be common consumer items and, while commercially available, require some training to use and interpret properly. In June 2011, Japan's Fukushima City—about ninety kilometers from the stricken nuclear power plant—announced the distribution of dosimeters to some thirty-four thousand children aged four to fifteen. The devices were to be read monthly in cooperation with medical institutions. A survey conducted in May 2011, in which 147 healthcare workers in areas near the plant wore "glass badge" (thermoluminescent) dosimeters, found exposures ranging between 0.1 and 0.7 mSv.

Researchers at Fukushima University came up with a different strategy for monitoring radiation levels deep in the region's mountain forests: fitting wild monkeys with collars containing dosimeters and GPS transmitters. Presumably, the monkeys will not be signing consent forms.

Effective dose

FROM A TIME at the beginning of the twentieth century when there was no way to measure radiation doses except by crude comparisons of how much damage they did to people's skin, to a hundred years later, when sophisticated dosimetry exists, but officials may be loath to gather the exposure data needed to make use of it in an emergency, the public has often had to fend for itself. When and if reliable information becomes available during a nuclear accident, for example, measurements may be expressed in a confusing constellation of ways that need careful sorting out.

The term *effective dose* refers to the amount of a particular type of radiation absorbed by a certain kind of living tissue, measured in SIEVERTS. Various "quality factors" have been determined over the years to rate the varying influence of ALPHA PARTICLES, BETA PARTICLES, X-RAYS, GAMMA RAYS, NEUTRONS, and so forth, and the vulnerability of different parts of the body. The effective dose of any exposure takes these factors into account and is the most accurate indication of the resultant health risk to an individual. Equal effective doses correspond to about the same overall risk, if age

Plastic humanoid shells used to test levels of radioactivity—
better them than us

and sex differences are not considered. For uniform "whole-body" exposures by a certain type of radiation, the effective dose equals the EQUIVALENT DOSE.

An older unit of measurement for effective dose, the rem (for "Roentgen equivalent man"), may still be found in government and industry documents, especially in the United States. One sievert equals 100 rem.

Enriched uranium

THE UBIQUITOUS ELEMENT URANIUM is found in natural ores that are mined commercially, such as pitchblende (uranium dioxide, UO_2), but its most useful radioisotope, URANIUM-235, makes up only about 0.7 percent of it. For most nuclear power reactors, the fuel is artificially enriched, starting with "yellowcake" (U_3O_8), to contain 3 to 5 percent uranium-235. This is called low-enriched uranium (LEU)—about 150 tons of it are needed per year, obtained from about five hundred times as much ore, to fuel a 1,000-megawatt commercial reactor. For nuclear weapons, high-enriched uranium (HEU) is required at "weapons-grade" concentrations of more than 80 percent uranium-235. HEU above 20 percent but below weapons-grade finds some use in experimental reactors—such as the Joyo "fast-neutron" reactor in Japan—and the manufacture of medical isotopes.

The process of raising the concentration of uranium-235 to useful levels is not easy, because uranium-235 is chemically identical to its far more abundant cousin, URANIUM-238 (which comprises about 99.3 percent of natural uranium, due to its 4.5-billion-year HALF-LIFE). Vast amounts of power are needed to run enrichment equipment. Huge by any industrial standard, the "gaseous diffusion" plants of the 1950s that pumped corrosive uranium hexafluoride gas through miles of pipes and membranes (the American facility at Oak Ridge, Tennessee, sprawled across two million square feet) were consuming 10 percent of the nation's electricity. In the 1960s, each plant devoured as much power per year as the city of New York. Modern plants, which use sophisticated gas centrifuges—whose rotors spin at 50,000 to 70,000 rpm—are more efficient, taking only about 5 percent as much electricity as diffusion, but the devices

Iranian president Mahmoud Ahmadinejad tours the Natanz uranium enrichment plant in 2008. In 1995, Iran received designs and components for 1970s-vintage centrifuges, which it has been seeking since then to improve, from Pakistani nuclear official Abdul Qadeer Khan. By 2012 it possessed about 6 tons of LEU and 220 pounds of HEU around 20 percent U-235.

themselves are one of the most guarded technologies in a world fearful of nuclear weapons proliferation. In 2009 and 2010, the United States and Israeli governments cooperated in a cyber-attack against Iran's uranium-enrichment plant, using a computer worm called Stuxnet to disable about a fifth of the country's working centrifuges.

Not that fear has prevented clandestine bomb development in the past. In 1976 the deputy director of the CIA briefed members of the U.S. Nuclear Regulatory Commission about the theft by the Israeli government of 100 kilograms of weapons-grade HEU from a U.S. Navy nuclear fuel plant in Pennsylvania.

Equivalent dose

ONE MIGHT SAY THAT this is the second-best measurement of human radiation exposure, after EFFECTIVE DOSE. Both are expressed in SIEVERTS, but equivalent dose takes into account only the type of radiation, not the particular organ or tissue exposed. In cases of "whole-body" exposure, where all cells are assumed to be uniformly irradiated, the equivalent dose is the average over the entire body and is the same as the effective dose.

Yes, these concepts are maddening, even to experts. If radiation were like a bullet, it would be easy to tell when someone got shot. But it is far subtler, though sometimes just as deadly.

Exposure

THE WORD *EXPOSURE* is of course used colloquially in myriad ways—with regard to liability law, accident insurance, public nudity—but it also happens to have a technical definition when the subject is radiation. One roentgen of exposure is equal to 258 microcoulombs per kilogram (μC/kg) of irradiated material. A coulomb, by the way, is a unit used to measure electric charge—named after the French physicist Charles-Augustin de Coulomb (1736–1806)—equal to the amount of charge carried by one ampere of electric current in one second. The ampere, by the way, is named after another French physicist, André-Marie Ampère (1775–1836), and is equal to 6.241×10^{18} electrons passing a given point in an electric circuit in one second (that is, one coulomb per second). Public nudity, by the way, is much more interesting than this.

Fallout

ETWEEN 1945 AND 1980, 502 nuclear weapons were detonated on or near the earth's surface (including 8 underwater) in tests of how they worked, with a total explosive energy equal to 440 million tons of TNT (189 megatons from FISSION bombs, 251 megatons from FUSION "H-bombs"). The debris thrown up into the atmosphere from this cold war frenzy, mostly in the Northern Hemisphere, still represents the largest collective dose of manmade radiation. After the Big Three nations (USA, USSR, UK) signed a partial test ban treaty in 1963, most explosions were confined underground, though thirty-eight of some nine hundred such American tests leaked into the outside air. The now-colloquial term *fallout* dates back to the 1945–50 period, stemming from early awareness that radioactivity could "rain" out of the sky at some distance from the bomb hypocenter—a fact dramatized by the deadly black rain that fell within twenty to thirty minutes after the Hiroshima blast. But as early as 1940, physicists Rudolf Peierls and Otto Frisch, who were working in England, wrote in a now-famous research memorandum about the possibility of atomic bombs that "owing to the spreading of radioactive substances with the wind, the bomb could probably not be used without killing large numbers of civilians, and this may make it unsuitable as a weapon for use by this country." Would that it had been so.

Many sins were committed during the cold war in the name of national security, but surely one of the most egregious was the campaign by the U.S. Atomic Energy Commission (AEC) during the 1950s to hide the dangers of fallout from the public. The history of this cover-up is by now well documented. During a 1979 congressional investigation, Peter Libassi, general counsel for the U.S. Department of Health, Education, and Welfare, testified that there was "a general atmosphere and attitude that the American people could not be trusted to deal with the uncertainties, and therefore the information was withheld from them. I think there was concern that the American people, given the facts, would not make the right risk-benefit judgments."

This sentiment was hauntingly similar to public assurances given by the Japanese government in the 1970s, when Japan was accelerating its reliance on nuclear energy, that it was completely safe to live near the reactors. The withholding of vital information by government and utility officials after the March 2011 Fukushima disaster showed the persistence of such attitudes at the highest levels.

Nineteen RADIONUCLIDES of significant concern to human health were produced and dispersed globally by the 1945–1980 weapons tests. About half had half-lives measured in days and were therefore troublesome mostly in the weeks and months following any test. Of the rest, PLUTONIUM-239, -240, and -241, CARBON-14, CESIUM-137, STRONTIUM-90, and TRITIUM are still very much with us. The estimated worldwide average dose to individuals peaked in 1963, at 0.11 mSv, and fell to below 0.005 mSv by the 2000s. Of course, these averages mask far higher personal exposures. For example, the highest effective doses downwind of the Nevada test site (eighty miles from Las Vegas, eighty-six detonations between 1951 and 1962) were perhaps 60 to 90 mSv, with thyroid doses ranging up to 4.6 Gy from the ingestion of IODINE-131 in milk. But all such historical figures—and subsequent epidemiological studies of related health effects—are compromised by the paucity of radiation measurements actually taken in the field at the time of the tests.

"Your best action is not be worried about fallout," advised a 1957 AEC booklet titled *Atomic Tests in Nevada*, the same year the agency began promoting atom bomb watching for tourists. "In the dawn's early light in the wake of a detonation, the atomic cloud can be seen attenuating across the sky," chimed reporter Gladwin Hill in the Sunday *New York Times* travel section. "It may come over an observer's head. There is virtually no danger from radioactive fallout."

The entire continental United States received fallout from the Nevada tests, with cumulative doses to individuals in the Midwest and East averaging 1 mSv. Hot spots occurred in unpredictable places far from the test sites, such as Albany, New York. During the years of active testing, ZIRCONIUM-95 was the main contributor to external exposure, but by 1966, cesium-137 had taken over this distinction. Children aged three months to five years received ten times the adult thyroid dose from iodine-131. Throughout the 1950s these dangers were known in international scientific circles, but were consistently denied by the AEC and routinely dismissed by mainstream press organs such as *Time* and *Newsweek*.

"Gee, that's important, isn't it?" President Eisenhower replied after his first full briefing on fallout. "Is everybody ready for lunch?"

The most horrific single fallout incident occurred in March 1954, when an American H-bomb was detonated above Bikini Atoll in the South Pacific. Radioactive ash from vaporized coral wafted down over an area of seven thousand square miles, inflicting disfiguring radiation burns ("beta burns") on Marshall Islanders. The average doses to the thyroid, due mainly to eating contaminated food, were estimated to be 12 Gy for adults, 22 Gy for children, and 52 Gy for infants. Cases of thyroid gland disorders, including cancer, began appearing ten years later, especially in children. A crew member of the Japanese fishing boat *Fukuryu Maru*, which had been motoring well outside the test zone, became the first atomic fatality since Nagasaki when he succumbed to radiation sickness nearly seven months later. The American government apologized and gave the man's widow $2,800. Instead of using the Western term *fallout*, the Japanese press called the phenomenon "ashes of death."

When the Soviet Union exploded the world's largest nuclear weapon in 1961—a 50-megaton fusion blast at an altitude of 3.5 kilometers, dubbed the "Tsar Bomb" in the West—they chose the remote island of Novaya Zemlya,

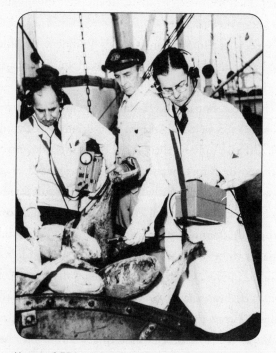

Hot tuna? FDA scientists use Geiger counters to check fish exposed to fallout from 1954 atomic bomb test in the Pacific.

north of the Arctic Circle in what is now the Russian Federation. Cesium-137 is still abundant in the region's lichens and reindeer meat.

The tenebrous era of atmospheric nuclear weapons testing is probably over, though there will always be some chance that a renegade state could ignore international outrage and openly detonate a bomb. The 1996 Comprehensive Nuclear Test Ban Treaty adopted by the United Nations General Assembly, which would outlaw *all* nuclear explosions, has been signed by 182 states and ratified by 155 (including Russia, but not the United States, which last tested in 1992). Even though a 2006 test by North Korea was conducted underground, it was confirmed by other governments through detection of radioactive debris, such as RADIOISOTOPES of XENON, in air samples, indicating leakage from the site. Fallout today is more likely to arise from nuclear reactor disasters, as experienced at Fukushima and Chernobyl. Since many of the same radionuclides are released by broken reactors, the terror of Nevada, Bikini, Novaya Zemlya, and myriad other former test sites around the world is nowadays visited upon populations who depend on these plants for everyday electric power. Concern focuses on iodine-131, cesium-134 and -137, strontium-90, and RUTHENIUM-103 and -106. In August 2011, inspectors in Ibaraki Prefecture found cesium-137 at 52 Bq/kg in rice from Hokota, about 155 kilometers (ninety miles) from the Fukushima plant. The Japanese regulatory limit is 500 Bq/kg (1,200 Bq/kg in the United States, where the average diet does not include as much rice). Rice production in Fukushima and neighboring Ibaraki and Miyagi prefectures accounts for 15 percent of Japan's total crop. Radiation above levels deemed "safe" had already been found in beef, spinach, lettuce, celery, broccoli, cauliflower, mushrooms, tea leaves, milk, and a plethora of other foods. For months after the Fukushima disaster, the Japanese government balked at placing comprehensive bans on food from the region.

As evidence that the global scope of fallout from stricken reactors is similar to that from nuclear weapons, sensors in the state of Maryland on the Atlantic coast of the United States detected small amounts (in femtocuries) of iodine-131 from Fukushima in the air at the end of March 2011. The radionuclide was also found in rainwater in Boise, Idaho, at 242 picocuries per liter, and in Massachusetts and Pennsylvania in the range of 40 to 100 picocuries per liter. Drinking water in Philadelphia held 2.2 pCi/L. Milk in Little Rock, Arkansas, held iodine-131 at 8.9 picocuries per liter, nearly three times higher than the U.S. Environmental Protection Agency (EPA) maximum contaminant level (3 picocuries per liter for persistent lifetime [seventy-year] exposure; the Food and Drug Administration limit is 4,700 picocuries per liter for a one-time exposure). Milk in Hilo, Hawaii, held 24 pCi/L of cesium-134, 19 of cesium-137, and 18 of iodine-131.

The EPA maintains a nationwide network of monitoring stations, called RadNet, to track radioactivity from nuclear accidents and weapons tests. The stations take regular samples of air, precipitation, drinking water, and milk. In the

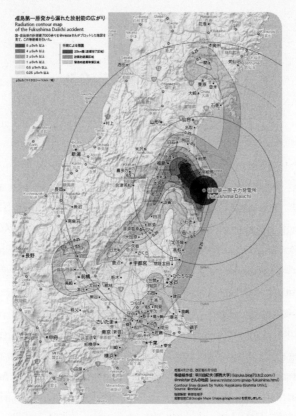

Map of fallout radiation levels after Fukushima Daiichi disaster

weeks after the Fukushima disaster, RadNet was able to detect concentrations of telltale radionuclides, such as the iodine-131 in rainwater in Boise, Idaho, on March 22, and 0.77 pCi/L in milk in Spokane, Washington, on March 25. There is no such thing as a local nuclear disaster.

Fission/Fusion

THE EVIL PRINCES of the nuclear kingdom—or the beautiful princesses, depending on your outlook. That their discovery and development coincided with World War II is one of the most regrettable coincidences in the history of science. At a less fearsome moment of the twentieth century, perhaps the headlong rush to make bombs could have been avoided.

The supreme achievement of the physics revolution that began in the last decade of the nineteenth century with the discovery of radioactivity was the detailed description of how matter is made of discrete particles. With Einstein's equation $E = mc^2$ came the realization that matter and energy are equivalent. When physicists realized by the late 1930s that an atom could be broken apart, producing fragments that did not have the same total mass as the original atom, the staggering amount of energy released in the form of GAMMA RAYS to make up the difference (keep in mind, that's the *speed of light squared*) astonished even the esoteric academics. But the practical reality of an atomic bomb remained obscure or was dismissed as infeasible by all but a few. Ernest

Rutherford, the field's patriarch, who first "split" the atom, died in 1937 thinking that the notion of exploiting nuclear energy was "the merest moonshine"—or perhaps just science fiction à la H. G. Wells's 1913 novel *The World Set Free*. Even in April 1945, four months before Hiroshima, the chemist and sociologist Michael Polanyi (1891–1976) and the philosopher Bertrand Russell (1872–1970) were asked on a British radio show what applications might come from $E = mc^2$, and they both drew a blank.

The word *fission* had been applied by biologists to the process of cell division since the mid-nineteenth century, but in physics, it dates to the momentous year 1939. It means the splitting of an atomic nucleus, usually with a NEUTRON, into—most often—two lighter nuclei. Certain heavy elements such as uranium produce more neutrons when they split, thus making a self-sustaining chain reaction possible as these fly away and strike other uranium atoms. The process can either be controlled in order to harness the energy given off during fission or allowed to cascade rapidly out of control to create a huge blast of energy. The former is a reactor, the latter a bomb. In both cases, the fission products are highly radioactive, requiring extensive CONTAINMENT in a reactor building or monitoring of FALLOUT after an open explosion.

Fusion is the reverse of fission, whereby two nuclei are forced together to form a heavier one. A great deal of energy is required to accomplish this, because all atomic nuclei are positively charged, and the repulsion of like charges must be overcome. In nature, this happens under the enormous

The human toll of nuclear fission: Hiroshima hospital, August 1945

gravitational compression at the center of stars. In a thermo-nuclear weapon (or hydrogen bomb), it happens when fission energy is used to fuse together deuterium (^{2}H) and TRITIUM (^{3}H) to create helium-4 (^{4}He) plus a neutron plus energy. So far, no one has figured out how to control fusion in order to harness the energy released. Fusion was first achieved in a laboratory in 1932; the first H-bomb was exploded twenty years later, by the United States.

Fusion is often thought to be "cleaner" than fission, because it does not create radioactive fragments. But the neutrons that are released can cause whatever they are absorbed by to become highly radioactive. It is not clean energy.

Francium-223

RANCIUM, THE QUINTESSENCE OF RARE, has no applications and you will never—well, hardly ever—encounter it in the environment. There are perhaps a few tens of grams of it in the entire bulk of the planet, found in vanishingly tiny traces in uranium ore. When scientists succeeded in collecting a few thousand *atoms* of it in 1996, they made news. It is exceedingly radioactive and decays into nasty stuff such as RADIUM and RADON, but this is one case where you can safely say, "So what?"

Francium's discoverer, in 1939, Marguerite Perey (1909–1975), was the first woman ever elected to the French Academy of Sciences. And she named it after *la belle France*. So she is honored here for carrying the torch of knowledge into fiercely misogynistic territory that had snubbed even her stellar mentor, Marie Curie. The academy was a gentlemen-only club from its founding in 1666 to Perey's admission in 1962.

Madame Curie hired Perey in 1929 to be her personal assistant—that is, her *préparatrice*, or lab tech, who performed the tedious chore of obtaining purified samples of radioactive substances for research at the Radium Institute in Paris. Curie must have known by then that this was dangerous work—the job was usually limited to three months, but Perey was so talented that she stayed for years. She soon suffered skin burns on her hands and fingers, absorbing radiation doses that would eventually kill her. Perey's account of her first meeting with Curie is a window into the latter's haunting presence just five years before her own death in 1934 from the effects of decades of radiation exposure:

"I was directed toward a small, very gloomy, waiting-room. Alone in this dismal atmosphere, the passing minutes seemed like hours to me. Then, without a sound someone entered like a shadow. It was a woman dressed entirely in black. She had grey hair, taken up in a bun, and wore thick glasses. She conveyed an impression of extreme frailty and paleness. At first I thought it was a secretary, but then I realized to my great embarrassment that I was in the presence of Marie Curie in person!"

Perey hoped that the short HALF-LIFE of francium (about twenty-two minutes for its least unstable RADIOISOTOPE, francium-233) would someday make it useful in cancer therapy, but the difficulty of obtaining even trace amounts of it made this impractical. In 1959, after many years of health problems, doctors found that Perey's bones emitted radiation characteristic of actinium, the element she had used to generate francium. From the early 1960s until her death, she suffered cruelly from radiation sickness and cancer—safety had never been a hallmark of the Radium Institute.

Francium was the last element discovered in nature—all the rest have been synthesized. Though appreciative of the personal satisfaction and public honors that came with discovering an element, Perey never forgot the "moments of tears and deceptions caused by vile traits of human character: manifestations of baseness and perfidy" that came with being a female scientist of her generation.

Gamma rays

GAMMA RAYS WERE THE THIRD principal type of radiation discovered and named at the scientifically fertile turn of the nineteenth into the twentieth century. Unlike ALPHA and BETA radiation, gamma rays are not composed of particles, but of massless packets of electromagnetic energy in extremely high-frequency waves (just like the photons of visible light, only at frequencies six orders of magnitude— and up—higher). Like X-RAYS, these waves are very penetrating. At the lofty reaches of their spectrum they can pass right through many kinds of matter as if it were transparent. When they interact with living tissue, they are harmful.

Paul Villard (1860–1934), the French chemist and physicist who discovered gamma rays in 1900 while studying radium, did not take advantage of the excelsior moment to name what he had found. He was a *lyceé* (high school) teacher hardly on the same rung of the scientific establishment as Ernest Rutherford, who in 1903 placed Villard's radiation into the Greek-letter rubric (α, β, γ) that Rutherford had already started.

Villard first recognized that gamma rays were different

from the X-rays that Wilhelm Roentgen had identified five years earlier, because of their greater penetrating depth. Since the dividing line between the two on the electromagnetic spectrum has never been precisely defined, the convention today is to call the electromagnetic radiation emitted by atomic nuclei gamma rays and by electrons outside the nucleus X-rays. Exceptions to this practice are COSMIC RAYS and the tremendously high-energy phenomenon in astronomy called the "gamma ray burst" (first noticed in the 1960s by American "Vela" satellites designed to detect clandestine nuclear weapons tests in space), which cannot be explained as radionuclide decay. The earth's atmosphere shields us from astronomical sources.

Gamma rays travel at the speed of light, so when they career through matter such as a living cell, the electrons they displace move so fast in turn that they cause few breakages, or ionizations, of surrounding molecules. This is known as "low-LET" radiation, where LET stands for linear energy transfer. Alpha particles, on the other hand, which are comparatively slow and massive, are termed "high-LET" and cause more damage for any given absorbed dose.

In scientific literature, "low-dose" exposures are defined as being from near zero to about 0.1 SIEVERTS (100 mSv) of low-LET radiation. But many radioactive sources give off a mixture of low- and high-LET. The average annual EFFECTIVE DOSE from natural BACKGROUND gamma rays is about 0.5 mSv, with maximums around 5 mSv.

In August 1945, hundreds of thousands of people

suffered very high gamma ray doses over short periods of time from the atomic bombs dropped on Hiroshima and Nagasaki by the U.S. Army Air Force. Doses decreased with distance from ground zero, with an average of about 300 mSv and maximums above 2,000 mSv to the colon. (An acute, whole-body gamma ray dose of about 1,000 mSv will, within one month, kill about half the people exposed.) With the priceless help of accurate records in the Japanese family registration system, or *koseki*, about 120,000 survivors—fewer than half of whom are still living—have been closely monitored since 1950 and comprise the single most important source of information for making risk assessments about the health effects of radiation. Of about ten thousand deaths due to solid cancers in this cohort, approximately 5 percent are attributable to radiation exposure. The death rate from leukemia is especially striking, with about 31 percent of all leukemia deaths due to exposure.

From 1945 to 1980, tens of thousands of military personnel were exposed mostly to gamma radiation during aboveground testing of nuclear weapons. About seventy thousand men from tests conducted by the United States and Great Britain form another important study group, though individual doses were mostly in the 1–4 mSv range. Studies of medical patients irradiated for treatment or diagnosis have also produced valuable data.

A measure of how well various substances will block gamma rays was developed by atomic weapons researchers in the 1950s. One "tenth-value thickness" lowers the radiation passing through the material by a factor of 10. Water thus has a tenth-value thickness of 24 inches; soil, 16 inches; concrete, 11 inches; steel, 3.3 inches. This practical-minded information was perhaps behind one of the most infamous official pronouncements of the cold war, when T. K. Jones, an undersecretary of defense in the Reagan administration, declared in 1981 that nations could survive nuclear war "if there are enough shovels to go around."

As the most energetic photons in the electromagnetic spectrum, gamma rays find many applications for their penetrating power. Gamma emitters—including CESIUM-137, COBALT-60, AMERICIUM-241, and TECHNETIUM-99—are the most widely used radiation sources in medicine and industry. There is no "pure" gamma emitter—most alpha and beta emitters also produce gamma rays.

Geiger counter

WHAT WOULD 1950S SCIENCE FICTION movies be without the Geiger counter's ominous clicking? Children of that era grew up knowing that if the clicks were fast and furious, you were in big trouble.

Invented by the German physicist Hans Wilhelm Geiger (1882–1945), these instruments detect radiation and give an indication of its intensity by registering the ALPHA or BETA particles or photons of GAMMA RAY energy being emitted. They are called "counters" because each particle produces an identical pulse—those clicks!—that can be tallied electronically. Nothing is determined, however, about the type or energy of what's causing the pulse. Hans Geiger did fundamental research with Ernest Rutherford on the structure of the atom, but pop culture images from the early Atomic Age made his handy gadget far more famous than his scientific work. Unlike many of the elite physicists of the prewar era, Geiger chose to remain in Germany during the Third Reich and participated in military efforts to exploit nuclear energy during World War II. Fortunately, Hitler never got anywhere on the road to owning an atomic bomb.

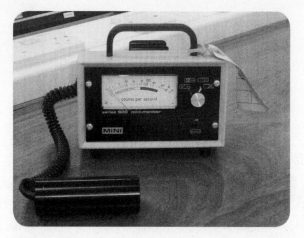

Radiation detection instrument and 1950s horror movie prop

Geiger counters have remained useful because they are simple and practical. They are commercially available to consumers, but are expensive and not effective for finding hazardous levels of radiation in food or water at home. (A manual from Japan's Ministry of Health, Labor, and Welfare advises against using Geiger counters for this purpose, because of their low sensitivity to gamma rays.) Unless you live near a nuclear power plant or like to spook your kids, save the money for a rainy day.

Gray

NAMED AFTER THE BRITISH PHYSICIST and radiologist Louis Harold Gray (1905–1965), who founded the field of radiation biology, or radiobiology, the gray (abbreviated Gy) is a unit of measurement for the amount of radiation absorbed by any exposed substance. It is defined as the absorption of one joule of energy from ionizing radiation by one kilogram of homogeneous matter. It is not related to the CURIE or BECQUEREL, which measure something completely different: the nuclear disintegrations in a source of radiation. So there is no conversion of the gray into those units or vice versa.

For biological tissues and organs, whose content is not strictly homogeneous, an absorbed dose of radiation expressed in grays is a rather crude average. Put bluntly, uncontrolled human exposures above 3 Gy can cause severe damage, leading to death in weeks or days. In order to get a more nuanced view of the risk of harmful effects, the dose has to be weighted according to the type of radiation and the kind of tissue exposed. The dose in grays is therefore multiplied by various factors, derived from laboratory or real-life experience, that reflect these differences. Once this is done, the weighted dose unit is called the SIEVERT.

Because the United States has clung to the nonmetric system of measurements, at least outside scientific circles, despite officially adopting metric in 1964, several old-fashioned units of measurement can still be encountered in government and industry documents. Before the gray was established in 1975, a similarly defined unit, called the rad (for "radiation absorbed dose"), which dated back to 1918, was used. Rad meant 100 ergs of energy per gram of matter. Since 1 erg equals 10^{-7} joules, this is the same as 0.01 joules per kilogram. Thus, 1 Gy equals 100 rad.

The gray is also used to measure an arcane dosimetric quantity called the kerma (for "kinetic energy released in matter"), defined as the initial kinetic energy of charged particles liberated by uncharged particles such as neutrons in a unit mass of material. This measurement is sometimes encountered in studies of Hiroshima and Nagasaki survivors. And as long as we're in arcane territory, we might as well deal with the roentgen (R)—a memorial to Wilhelm Roentgen (1845–1923), the discoverer of X-RAYS—which is a unit of exposure to X-rays and GAMMA RAYS equal to the amount of those radiations that will produce a certain electric charge in dry air. An exposure of 1 R is approximately equal to 10 mGy in soft tissue.

Is it any wonder that the public becomes confused when these terms are tossed around by experts, let alone in the news media whenever there is a nuclear disaster? Technical jargon is useful for scientists and engineers, who spend many years learning how to communicate with one another in precise terms. For laymen, however, such specialized language acts as a barrier that can, unfortunately, be manipulated by the unscrupulous to maintain a fog of ignorance.

Half-life

HALF-LIFE IS ONE THOSE RARE bits of scientific jargon whose meaning is literal. As a RADIONUCLIDE decays, its half-life is the amount of time that passes before half of it is gone. Why is this useful to know? It is simply the most accurate means of identifying the thousands of radioactive isotopes.

In the first years of the twentieth century, physicists struggled to understand many confusing puzzles about radioactivity. One of the thorniest was that uranium, radium, and a few other elements showed the same level of activity whenever they were measured, but polonium and RADON (called "emanation" at this early juncture) rose and fell over time. Yet the activity of any given radioactive element always decreased by half in the same period of time. Organizing these observations into a comprehensive theory of radioactivity absorbed the sharpest minds in physics and chemistry. It was the great Ernest Rutherford who finally saw that each radioactive element was "genetically" related to others— that is, they were the progeny (or *daughters*, as the old-fashioned terminology put it) of certain parents and would die at fixed rates indicated by their half-lives. It helps to remember that this was still an era when the transformation of matter reeked of alchemy.

Half-lives stretch from the unimaginably brief to the unimaginably long. For example, beryllium-8 has a half-life of 7×10^{-17} seconds, while tellurium-128 clocks out at 2.2×10^{24} years. The shorter the half-life, the more radioactive. A handy-dandy formula for figuring out the fraction of a radioactive substance that remains after a certain number n of half-lives is $\frac{1}{2}^n$. If you can wait long enough, all radioactivity goes away.

Biological half-life is the time required for biological systems to eliminate one-half of a radionuclide harbored by the body. For example, the whole body biological half-life of POLONIUM-210, whose physical half-life is 140 days, is about 50 days. The *effective* half-life of Po-210 (the time needed for the combined action of the physical and biological half-lives to reduce the activity by 50 percent) is approximately 40 days.

Iodine-123, -124, -125, -131

WHEN RADIOISOTOPES OF IODINE get inside you, they congregate in your thyroid gland—a little blob shaped like a butterfly right below your Adam's apple—which needs natural iodine to make hormones for good health. Unfortunately, it cannot tell the difference between radioactive and stable forms of the element. This makes "radioiodine" a potent carcinogen when it runs wild, but potentially beneficial in controlled doses. Thyroid cancer is a major concern associated with accidental exposures, especially in children and adolescents. On the flip side, medical doses have been used for more than half a century to help cure thyroid disorders.

There are more than three dozen iodine RADIONUCLIDES, but most have half-lives that are far too short to be of practical value. Some are troublesome as short-lived FISSION products produced by nuclear reactors and bombs. (FYI: reactors and bombs produce essentially the same fission products, but the extremely fast time scale of a bomb explosion spawns a different mix.) All are powerful emitters of BETA PARTICLES. If released into the environment, they can be absorbed by plants and enter the food chain. Historically, the worst human exposures have come from drinking milk produced by cows that grazed on contaminated pastures. Because kids drink lots of milk, they take the hardest hit.

Some 2.3 million children who lived near the burned-out Chernobyl nuclear power plant—in Belarus, Ukraine, and eastern parts of the Russian Federation—in 1986 have experienced regional increases as high as 100-fold in the normal incidence of thyroid cancer. Thousands suffered thyroid doses of several GRAY, but most who have developed cancer were exposed to less than 300 mGy. The spike in cases has been documented as far as five hundred kilometers from the reactor site. Nuclear catastrophes always travel far and wide.

Iodine-131 happens to be generated in prodigious quantities by nuclear reactors and bombs. (Iodine-132 and -133 are produced in lesser amounts.) Along with CESIUM-137, it is the greatest threat after a nuclear power accident. When it escapes, it disperses rapidly in gaseous form and falls to earth in rain. With a HALF-LIFE of eight days, only about seven one-hundredths of a given amount of it is left after a month, so when it gets into the biosphere, it brings a large initial wave of exposure that dissipates rather quickly. This helps explain why children under ten years old at the time of the Chernobyl disaster suffered the most dramatic increases in thyroid cancer starting about five years later, while those born after 1986 were largely spared. Risk is heavily dependent on age at exposure—highest for infants and small children, relatively low after the age of twenty, and almost zero after forty. Some 360,000 Japanese children who were eighteen or under at the time of the Fukushima Daiichi reactor disaster will be given ultrasonic thyroid exams every two years until they turn twenty.

The U.S. Food and Drug Administration's limit for iodine-131 in milk is 4,700 picocuries per liter for acute exposures. That is, milk above this level should not be drunk even once, and must be removed from the market. The U.S. Environmental Protection Agency, which does not regulate milk, sets an annual exposure limit for drinking water at 700 pCi/L. The EPA lifetime (seventy years) limit is 3 pCi/L. Why the differences? The scientific rationale is that long, low-level doses are less dangerous than short, high-level ones. The realpolitik explanation is that drinking water can be cleaned rather cheaply, but dumped milk is a big loss for dairy farmers. During the weeks after the Fukushima disaster, milk samples from Los Angeles and Phoenix contained iodine-131 at the EPA lifetime limit for water. A Little Rock, Arkansas, sample topped the charts at 8.9 pCi/L. But they were not declared a threat to public health.

Such exposure limits are the result of many years of research, including human subject experimentation that clearly broke the boundaries of ethical conduct. In 1995 the U.S. Department of Energy began publishing a summary of human radiation experiments carried out from the early

1940s through the 1970s. Typical is the iodine-131 trial as described in the report:

INGESTION OF MILK CONTAINING IODINE-131

THIS STUDY* was conducted in 1963 by a graduate student at the University of Rochester to investigate the body's metabolism of radioiodine found in dairy products. The research sought to determine whether iodine found in milk was transferred to the thyroid in the same quantities as the inorganic iodide commonly used in medical studies. As much as 40 percent of the iodine found in milk was found to be protein bound. The study focused on the range of uptake percentages in children of various ages.

Subjects for the experiment were chosen with an emphasis on the younger age groups, since the majority of known research had been conducted on adults. The subjects ranged in age from 6 years to 50 years; seven were less than 21 years old. The milk used for this study was obtained from Cornell University's Department of Veterinary Medicine, where a cow had been fed iodine-131 so as to produce 5 to 10 nanocuries per liter in its milk.

---

* R. G. Cuddihy, "A Hazard to Man from I[131] in the Environment," *Health Physics* 12 (1966): 1,021–25.

All subjects were put on an iodine-restricted diet
prior to the study and then were fed the iodine-131
milk for a minimum of 14 days. One of the children in
this study subsequently developed a benign nodular
hyperplasia of the thyroid, which was later surgically
removed. The research was performed under a contract with the U.S. Atomic Energy Commission.

Less than twenty years after the Nuremberg trials of
Nazi war criminals, scientists at an elite American university
apparently felt free to conduct these tests on children.

A healthy thyroid gland absorbs about 20 to 30 percent
of ingested iodine-131, while someone with hyperthyroid-
ism can soak up 60 percent. If a stable form of iodine can be
given to people before or immediately after exposure (within
four hours), however, their thyroids will become saturated
with it and will not take on any new iodine that is radioactive. For this reason, potassium iodide is the first line of defense against iodine-131 when nuclear disasters occur. It is
not 100 percent effective, but it is very good. The World
Health Organization recommends that children take potassium iodide if exposure levels reach 10 mGy, and that everyone else take it at 100 mGy (with the exception of people
over forty years of age, who really need it only at very high
levels, around 5 Gy, to prevent acute effects). Of course,
exposure figures are often not known while disasters are
happening, so the safe side is for everyone to take it on a
daily basis until the danger is gone. Long-term dosage is

not recommended, because of adverse side effects, and the correct dose is lower for children than for adults. (Newborns less than a month old, and pregnant or breastfeeding women, should not take it more than once.) Ideally, emergency management officials instruct exposed populations on how to use the nonprescription tablets. Regions of the former Soviet Union were especially cursed with thyroid cancer after Chernobyl because the government did not distribute any potassium iodide. Poland, on the other hand, provided some seventeen million pills to its citizens—ten million to children—and thereby dodged a heartbreaking epidemic.

Iodized table salt is of no help whatsoever in a nuclear disaster scenario. It simply does not contain enough iodine. This did not stop panicky consumers along China's coastline from ransacking grocery stores after the Fukushima debacle.

Because radioiodine is a dissolved gas in contaminated water, cleaning it out requires a more sophisticated filter than just the "reverse osmosis" membranes recommended by the EPA for most pollutants. A three-way device that combines activated charcoal, ion exchange (used in water softeners), and reverse osmosis is the best solution for tap water. Needless to say, if your drinking water is contaminated with radioiodine to an extent that requires filtration, you should evacuate the area.

In early November 2011, traces of iodine-131 (from single-digit to as high as 65 $\mu Bq/m^3$) were detected in the air across Austria, Czech Republic, France, Germany, Poland,

and Slovakia, creating fear of an unknown or unannounced reactor leak or other radiation accident somewhere in the region. In mid-November, the Hungarian Atomic Energy Authority revealed that the Institute of Isotopes Ltd., which produces radioisotopes in Budapest for health-care, industrial, and research applications, had released iodine-131 into the atmosphere in September and October. An institute official at first denied the charge, but its CEO soon admitted that though there had "not been any extraordinary event," a "certain amount is always emitted" and "in this case, however, it was higher than expected."

Now the good side. Iodine-131, with its energetic beta particles, is used in high doses to destroy abnormal thyroid tissue such as nodules, and malignant cells. Various other radioisotopes are chosen for different medical tasks: iodine-123 for brain, thyroid, and renal imaging; iodine-125 for brain, blood, prostate, and metabolic function diagnostics or therapy; iodine-132 for brain, pulmonary, and thyroid diagnostics. It all depends on the half-life and how much beta (and accompanying GAMMA) radiation is desired. For example, radiation from iodine-131 is about 85 percent beta and 15 percent gamma.

Ever since the days of professors Becquerel and Roentgen, radiation has been a double-edged sword. Radioiodine is the quintessential example.

Iridium-192

THERE ARE MORE THAN 30 RADIOISOTOPES of the element iridium, which was discovered and named in 1803 after millennia of being crafted in objects made of native platinum, its close chemical relative. The English chemist Smithson Tennant (1761–1815) persevered where others lost patience, identifying iridium in the residue left from refining platinum ores in his laboratory. He found beauty in the colorful iridium salts he then concocted, naming the element after the Greek goddess of the rainbow, Iris.

Despite its being silvery and hard and even more tarnish-resistant than platinum, no one would want to wear jewelry containing iridium-192. It is a strong emitter of GAMMA RAYS, and of BETA PARTICLES, which makes it useful only as a source for industrial radiography and medical radiotherapy. It is entirely man-made, usually by bombarding pieces of iridium metal with NEUTRONS in a reactor. It is not a FISSION product, so it is not found in FALLOUT from old nuclear weapons tests. Like other radioisotopes that have wide commercial use, such as COBALT-60, iridium-192 has starred in myriad accident stories in which sealed discs, capsules, or pellets are lost or stolen and wind up causing multiple cases of ACUTE RADIATION SYNDROME (ARS) among unsuspecting victims.

Typical in ignorance though extraordinary in casualties was an incident in Morocco in 1984, when a worker stole a 6×10^{11} BECQUEREL iridium-192 source from equipment used to inspect welding joints. He took the source home and kept it for eighty days (iridium-192's HALF-LIFE is 74 days) on a shelf in a bedroom where he, his wife, and four children slept. He died from ARS forty-four days later. His wife and children were dead within a month after that. A cousin and his mother also died after visiting the house. Two grandparents and another cousin who stayed there during the father's illness were finally evacuated to the Curie Institute in Paris, where they were treated and survived. There seems to be something mesmerizing and irresistible about the invisible power of radiation, known to the scientists who discovered it and experienced by the uneducated ever since.

Iron-55, -59

THERE IS MORE IRON than anything else on earth. Every living thing contains it. We need it for everyday good health—about two-thirds of the body's iron is in the blood. Fortunately, most of its two dozen RADIOISOTOPES have short half-lives and are not of environmental concern. But one, iron-55, is produced when NEUTRONS bombard the two most common natural iron isotopes, iron-54 and -56 (which are not radioactive). It was therefore produced in abundance—about 1,530 peta-BECQUERELS—by atmospheric nuclear weapons tests during the cold war era, though its 2.7-year HALF-LIFE made it a short-term hazard only. In 1965, iron-55 was detected in blood samples from Native Americans and caribou around Anaktuvuk, Alaska, at concentrations eight times higher than in samples taken from residents and beef cattle in Washington State, due to the accumulation of iron-55 from bomb FALLOUT on lichens. In the 1960s, iron-55 was also found to be one of the most widespread radioactive contaminants of marine life, such as salmon and cod, especially in the Northern Hemisphere.

Iron-59, with a half-life of 44.5 days, can be used as a tracer in studies of the blood and iron metabolism. It has been associated with increased risk of prostate cancer in nuclear workers.

Krypton-81, -85

KRYPTOS MEANS "HIDDEN" IN GREEK, which appealed to the British chemist William Ramsay (1852–1916) in 1898 when he found this rare element while searching for noble (or inert) gases to fill empty boxes in the periodic table. He put his schoolboy classical education to good use when he also discovered ARGON (meaning "the lazy one" in Greek, because it would not react chemically with anything else), neon ("new"), and XENON ("strange"). He wanted to call a certain phosphorescent gas niton (from the Latin *nitere*, meaning "to shine"), but the perfectly fine name did not stick, being dropped later in favor of RADON. Ramsay won the 1904 Nobel Prize in chemistry, but unfortunately did not prevent the eventual obsolescence of classical education.

Krypton has dozens of RADIOISOTOPES, most with very short half-lives and therefore of little environmental concern. The two exceptions are krypton-81 (229,000 years) and krypton-85 (10.78 years), both of which are produced when COSMIC RAY neutrons collide with stable krypton gas in the upper atmosphere. Krypton-85 is also a FISSION product of URANIUM-235, which means that it was spewed into the atmosphere by cold war nuclear weapons tests (about 185 peta-BECQUERELS) and is a common pollutant from nuclear reactors. About 1.85 PBq was released by the rupture of fuel rods during the 1979 Three Mile Island reactor accident in Pennsylvania. Even during normal reprocessing of spent fuel, a large nuclear power plant releases about 11 PBq of krypton-85. As a gas, it spreads easily, like a cloud, but because it is inert, it does not take part in body chemistry if ingested. Its BETA PARTICLE radiation is therefore considered primarily an external hazard to the skin. Krypton-81, however, emits GAMMA RAYS that can irradiate the entire body.

Krypton-81 is used to track the movement of underground water. Because it is generated only in the atmosphere, it is not found underground unless it was carried there by water. When surface water goes underground, it stops picking up krypton-81 from the air. With the knowledge of how slowly krypton-81 decays, then counting the remaining radioisotopes in a sample of aquifer water, you can calculate how much time has passed since the water was above ground. This method can look back much further than similar CARBON-14 dating, thanks to krypton-81's much longer HALF-LIFE. The trick is to isolate and count the exceedingly rare radioisotopes, a feat accomplished by illuminating them with lasers whose frequency matches the atoms' natural vibration, thus causing a resonant excitation that shows up as a bright spot on a camera. The million-year-old Nubian Aquifer's flow across northern Africa has been analyzed in this way.

In August 2011, krypton-85 was detected in air samples from the damaged reactor Number 2 at Fukushima Daiichi in Japan. It would not have robbed Superman of his powers, which were sorely needed by the hapless Japanese utility company TEPCO.

Lanthanum-140

THE DOZENS OF RADIOISOTOPES of the element lanthanum are mostly ignorable, because their half-lives are either so long that they are essentially stable (i.e., the primordial lanthanum-138, 102 billion years), or so short that they disappear in the blink of an eye. Lanthanum-140, a BETA PARTICLE emitter at 1.68 days, is notable because it is a FISSION product, the daughter of BARIUM-140. During the cold war era of atmospheric nuclear weapons tests, it was detected in snow in Central Europe after Soviet detonations in the autumn of 1961. At the end of March 2011 it was found in stagnant water on the basement floor of the turbine building of Reactor 1 at the crippled Fukushima Daiichi nuclear power plant, at 340 BECQUERELS per cubic centimeter.

Lanthanum has the distinction of having played a supporting role in the discovery of nuclear fission. In 1937, Irène Joliot-Curie (1897–1956)—the daughter of Pierre and Marie Curie—bombarded uranium with neutrons and detected a radioactive substance with a HALF-LIFE estimated to be 3.5

hours that behaved chemically like lanthanum—she thought it was a new element, but it was actually lanthanum-141. In Germany, the experiment was repeated and confirmed by radiochemists Otto Hahn (1879–1968) and Fritz Strassmann (1902–1980), because they had expected such neutron irradiation to produce elements that were heavier than uranium, not lighter. The momentous explanation was provided in early 1939 by Austrian physicists Lise Meitner (1878–1968)—Hahn's longtime research partner, who had fled the Third Reich—and Otto Frisch (1904–1979), who called the process fission and pointed out the enormous energy released as the missing mass was converted according to $E = mc^2$. Then Hitler invaded Poland, the United States launched the Manhattan Project, and the world was never the same again. Semiretired in neutral Sweden, Meitner did not learn of the Bomb until it was dropped on Hiroshima. "You must not blame us scientists for the uses to which war technicians have put our discoveries," she said. "My hope is that the atomic bomb will make humanity realize that we must, once and for all, finish with war." Hahn won the 1944 Nobel Prize in chemistry for his part in discovering fission—he could not attend the ceremony in December 1945 because he was interned in England with other German nuclear scientists—but Meitner went unrecognized. Albert Einstein fondly called her "the German Marie Curie."

Like her mother, Irène Joliot-Curie won a Nobel Prize

in chemistry—for the discovery of artificial radiation, which opened the door for man-made RADIONUCLIDES in medicine—in 1935, with her husband, Frédéric Joliot (1900–1958). Like her mother, she died of leukemia after years of radiation exposure in the laboratory.

Lead-210, -212, -214

O F THE DOZENS of lead RADIOISOTOPES, these three are notable because of their association with dangerous RADON gas. Lead-210 and -214 are DECAY PRODUCTS of radon-222, and lead-212 is spawned by radon-220 (thoron). From the approximately 4.3 EXABECQUERELS (4.3×10^{18} Bq, or 4.3 Ebq) of radon-222 released from the earth's surface every year, about 25.5 peta-Bq (25.5×10^{15} Bq, or 25.5 PBq) of lead-210 is formed. The atmosphere also holds a small amount of man-made lead-210 from cold war nuclear weapons tests and nuclear reactors. For example, the activity of lead-210 originating from bombs in 1962 was about 33 PBq.

Each of the three emits BETA PARTICLES, with half-lives of 22 years, 10.6 hours, and 27 minutes, respectively. Because of its longer HALF-LIFE and membership in the notorious decay chain of URANIUM-238, trace amounts of lead-210 are everywhere and part of natural BACKGROUND RADIATION. But above-normal concentrations are found in and around uranium mines, creating an occupational hazard. Like POLONIUM-210, lead-210 gets into tobacco leaves and then into smokers, whose rib bones have been found to contain twice as much lead-210 as those of nonsmokers.

Samples of bone and teeth from reindeer in the western Russian arctic region that lived before, during, and after Soviet nuclear weapons testing on Novaya Zemlya island have been found to contain elevated lead-210. Concentrations in bone ranged from 0.20 to 1.45 and 0.40 to 0.84 Bq per gram in animals that lived during and after the bomb tests, respectively. The level in a mixed sample of enamel and dentin from reindeer that died in 1890 was 0.025 Bq/g. No wonder Rudolph's nose glows.

Linear no-threshold (LNT)

A NY DOSE OF RADIATION greater than zero is potentially harmful. Many otherwise smart scientists and doctors have been dragged kicking and screaming into this consensus. There are still those who speak of "safe" levels of radiation, usually meaning exposures so small that no one need worry about them. For example, the U.S. Nuclear Regulatory Commission (NRC) states on its website that "a yearly dose of 620 millirem [6.2 mSv] from all radiation sources has not been shown to cause humans any harm." Yes, there are very low doses of natural BACKGROUND RADIATION that are inescapable, so fretting about them is perhaps a waste of time. But even these are not "safe" in the sense of being risk-free. And the standard retort that aspirin is not risk-free, either, is philistinism.

The "no dose is safe" consensus is expressed by a mathematical "risk model" called linear no-threshold (abbreviated LNT). In 2006 the U.S. National Research Council released the seventh in a series of periodic reviews of the Biological Effects of Ionizing Radiation (commonly referred to as the BEIR VII report) that began in 1972. The report stated that the LNT model "provided the most reasonable description of the relation between low-dose exposure to ionizing radiation and the incidence of solid cancers." While conceding that at doses below 100 mSv, or about forty times the average yearly background level, statistical limitations make it "difficult to evaluate cancer risk in humans," a "comprehensive review of the biology data" led to the firm conclusion that "risk would continue in a linear [straight-line] fashion at lower doses without a threshold and that the smallest dose has the potential to cause a small increase in risk." Worth repeating: *The smallest dose has the potential to cause a small increase in risk.*

To illustrate the LNT model, the BEIR VII committee presented an example of the expected cancer risk from a one-time exposure of 100 mSv above background. The odds depend on both sex and age at the moment of exposure, with women and children at higher risk, but the LNT model predicts that about one person out of a hundred will develop solid cancer or leukemia during his lifetime (forty-two will get cancer from other causes). From an exposure of 10 mSv, about one person out of a thousand will be stricken. Lifetime (defined as seventy years) exposure *just to natural background radiation* will cause one person out of a hundred to get cancer—and this calculation excludes exposure to RADON and other high-LET sources. The committee called special attention to the Oxford Survey of Childhood Cancer, which found a 40 percent increase in the cancer rate among children up to age fifteen who suffered radiation doses of 10 to 20 mSv in utero or during early life.

Addressing directly the school of thought that says "risks are lower than that predicted by the LNT, that they are nonexistent, or that low doses of radiation may even be beneficial," BEIR VII stated flatly that after reviewing such articles, the committee "does not accept this hypothesis," which is based on "findings not representative of the overall body of data." It would seem, then, that the NRC is quite fond of those nonrepresentative findings.

Such is the diplomatic language of a scientific smack-down. There is no such thing as a safe dose of radiation. One vocal member of the fringe movement known as hormesis, which holds that exposure to environmental toxins may be good for health, is Edward J. Calabrese, a professor of public health at the University of Massachusetts. Calabrese has attacked the BEIR consensus and the LNT model, going so far as to call Hermann Muller—who won the 1946 Nobel Prize in medicine for showing that X-RAYS can induce genetic mutations and was an early advocate of the no-safe-dose hypothesis—a liar. In 2009, Calabrese received what is perhaps the world's most cynically named award, from the Paris-based pro-nuclear industry lobby called World Council of Nuclear Workers: the Marie Curie Prize for "outstanding achievements in research on the effects of low and very low doses of ionizing radiation on human health."

Medical radiation

THE USE OF RADIATION in medical procedures has a long history that mixes the brilliant with the sordid. For the first three-quarters of the twentieth century, the bad effects of radiation were often ridiculed or ignored in the name of demonstrated or merely wished-for benefits. Regulatory standards have since become more and more strict. Physicians today are taught that exposure must be limited to the lowest practicable levels and subject to careful risk-versus-benefit analysis, but horror stories still occur.

After Wilhelm Roentgen announced his discovery of X-RAYS in January 1896, within months they were being used around the world to locate bullets in gunshot wounds, set broken bones, kill diphtheria germs, treat cancer, and diagnose tuberculosis, pneumonia, and enlarged hearts and spleens. In May of that same year, Thomas Edison gave the first major public demonstration in the United States for crowds of people who lined up for a chance to "see their bones." Also within months, reports began to appear in the scientific literature about painful and slow-healing irritation of skin exposed to Roentgen's rays. The death of Edison's assistant, Clarence Dally, from skin cancer in 1904 was the first mortality associated with man-made radiation. But it did not matter. In 1904, Dr. W. J. Morton (1846–1920), a prominent New York neurologist and member of the New York Academy of Medicine, promoted a quack radium elixir he called Liquid Sunshine, with which "it will be possible to bathe a patient's entire interior in violet or ultraviolet light." In 1911, eager experimenters at the University of Pennsylvania tried to take X-ray photographs of the human soul. The medical profession was especially enthusiastic about using X-rays and radium on women to treat both cancer and a host of non-life-threatening gynecological problems. Yet it was often doctors themselves who experienced severe radiation burns as they used their bare hands to calibrate primitive X-ray equipment. The foremost X-ray practitioner in the United States, Dr. Mihran Kassabian (1870–1910), warned fellow physicians not even to use the word *burns* to describe damage from overexposure, lest development of the new miracle tool be inhibited. His own hands were mutilated by X-ray exposure, and he died of raging skin cancer.

Today, medical use of radiation usually occurs under three circumstances: diagnostic examinations, generally via X-ray images; treatment of malignant disease (known as radiation oncology), usually by targeted external exposure to COBALT-60; and treatment of benign disease, which was largely phased out in the 1970s. There is also so-called nuclear medicine, in which a radioactive pharmaceutical is put into the body to produce images of certain organs or structures. About 400 million diagnostic medical exams and 150 million dental X-ray exams are performed annually in the United States, whose citizens receive the most such exposure in the world, amounting to an average yearly dose of 0.5 millisieverts per person. When all other medical uses of radiation are averaged in, the per capita dose rises to 6.2 mSv. By comparison, the average dose of natural BACKGROUND RADIATION in the United States is 2.95 mSv. Worldwide, about 3.6 *billion* diagnostic X-ray exams are conducted every year. Since they were introduced in the 1970s, increasingly popular computed tomography (CT) scans, which produce detailed three-dimensional images and are used even to browse for signs of disease among people with no symptoms, result in much higher doses, on the order of tens of millisieverts per exam. Cumulative doses of 100 millisieverts for children are not unusual.

About 50 percent of cancer patients are treated with radiation, in most cases receiving high doses of 40 to 60 GRAY to a small region of the body, with the aim of killing cells. Cervical cancer patients may receive doses as high as several hundred gray, significantly increasing their risk for later cancers of the bladder, rectum, stomach, and vagina, and for leukemia. For treatment of childhood cancers, the risk for developing a second cancer in the twenty-five years after diagnosis of the first is as high as 12 percent. Of course, many patients who endure radiation therapy are not being treated with cures in mind, but rather for palliative reasons.

Simply because they will not live very long, they are at little risk for subsequent radiation-induced malignancies.

In practice, the potential for errors and accidents—everything from miscalibration of beam machines to placing someone feetfirst into a CT scanner rather than headfirst—leading to injury of patients and staff is significant, as shown by the thousands of mistakes archived on the website of the Radiation Oncology Safety Information System (ROSIS). Needless to say, without radiotherapy, myriad patients would not survive their cancer at all.

How much radiation is risky? The best estimates are based on the 1986 Chernobyl nuclear power plant accident and studies of Japanese atomic bomb survivors who had excess cancer risk after exposures of 50 to 150 mSv. A chest or abdominal CT scan involves 10 to 20 mSv, versus 0.01 to 0.1 for an ordinary chest X-ray, less than 1 mSv for a mammogram, and as little as 0.005 for a dental X-ray. A study by Columbia University researchers, published in 2007, estimated that in a few decades, as many as 2 percent of all cancers in the United States might be due to radiation from CT scans given today. Because previous studies suggest that a third of all tests are unnecessary, the researchers concluded that twenty million adults and more than a million children are being put needlessly at risk. Yet about seventy-five thousand chest patients endured the dubious practice of double-scanning in 2008, the most recent year for which figures are available.

Which tests are overused? The International Commission on Radiological Protection cites routine chest X-rays when people are admitted to a hospital or before surgery; imaging tests on car-crash victims who do not show outward signs of head or abdominal injuries; and low-back X-rays in older people with degenerative but stable spine conditions. Even when tests are justified, they often include more views than needed and too much radiation. The top offender is the chest CT scan looking for clogged arteries and heart problems.

The U.S. Food and Drug Administration announced in 2010 that it would begin regulating medical radiation in the hope of reducing unnecessary exposure from three

only x-ray-
radium-
surgery-
ever cured CANCER
UNITED STATES PUBLIC HEALTH SERVICE

U.S. Public Health Service quackery prevention poster

imaging procedures: CT scans; nuclear medicine studies, in which patients are given a radioactive substance and doctors watch it move through the body; and fluoroscopies, in which a radiation-emitting device provides a continuous internal image on a monitor. "These types of imaging exams . . . can increase a person's lifetime cancer risk," the FDA said. "Accidental exposure to very high amounts of radiation also can cause injuries, such as skin burns, hair loss, and cataracts."

Even though medical radiation is better managed today than ever, patients receive far more radiation than ever before. The average lifetime dose of diagnostic radiation—excluding therapeutic radiation—has increased sevenfold since 1980, prompting widespread concern that certain procedures are overused and that they needlessly put patients at an increased risk of cancer. Children and women are particularly vulnerable.

Never be afraid to ask your doctor questions about the management of medical radiation.

Molybdenum-99

SOME PARENTS LIVE in the glory of their children. Molybdenum-99 is notable because of its DECAY PRODUCT daughter, TECHNETIUM-99M, the most widely used RADIOISOTOPE in nuclear medicine. Molybdenum-99's HALF-LIFE of about sixty-six hours is perfect for making portable technetium generators, sometimes called cows, which are delivered weekly to hospitals, where they are chemically milked (the technical term is *eluted*) for technetium-99m, to be used in diagnostic imaging tracers.

As a FISSION product of URANIUM-235, molybdenum-99 is created by bombarding ENRICHED URANIUM targets with NEUTRONS in nuclear reactors. Other medical isotopes, such as IODINE-131 and XENON-133, are by-products of this process. There are currently no producers of molybdenum-99 for medical use in the United States. Until 1989, Cintichem Inc. produced it for the American domestic market using a research reactor in Tuxedo, New York, but the plant was shut down after a TRITIUM leak contaminated water near the site. The U.S. government exports highly enriched uranium (HEU) originally made for nuclear weapons—uranium-235 at concentrations of around 93 percent—for manufacturing more than 95 percent of molybdenum-99 around the world through the Department of Energy or in cooperation with the European Atomic Energy Community (EURATOM).

A 2011 investigation by the U.S. Government Accountability Office found alarming deficiencies in monitoring the whereabouts of this weapon-grade material, with auditors unable even to verify the location of 1,160 kilograms of the 17,500-kilogram inventory. Its attractiveness to terrorists or rogue states is clear from the fact that the HEU-based bomb dropped on Hiroshima in 1945 contained about 64 kilograms of HEU, with an average enrichment of 80 percent. A well-designed weapon of mass destruction could work with less than 25 kilograms. Because about 97 percent of the uranium irradiated to obtain molybdenum-99 ends up as waste, tens of kilograms of this discarded HEU accumulate every year worldwide—somewhere.

Because of intense demand from the medical industry for molybdenum-99 generators of technetium—since 2009, shortages linked to reactor shutdowns among foreign suppliers have disrupted reliable shipments to hospitals—several commercial companies are seeking permits to run production reactors in the United States that would employ low-enriched uranium (LEU) that cannot stoke nuclear weapons.

Neptunium-237, -239

NEPTUNIUM IS A JUNK ELEMENT of the nuclear age. Artificially created in 1940 in a cyclotron at the University of California, Berkeley, by bombarding URANIUM-238 with NEUTRONS, neptunium-239 was named after the next planet beyond Uranus, the namesake of uranium, its next-door neighbor in the periodic table. This is as poetic as anything about neptunium ever gets, though it was once called bohemium.

Neptunium is basically a useless, radiotoxic by-product of PLUTONIUM production. Manhattan Project scientists looking for elements that could FISSION and thereby destroy cities tossed neptunium aside in favor of plutonium. (Obviously they were into the planets—Pluto is next out from Neptune, as plutonium follows neptunium in the periodic table.) Six decades later, their descendants at the University of California's Los Alamos laboratory were still at it, calculating that about 60 kilograms of neptunium-237—the most common and stable isotope, an ALPHA PARTICLE emitter with a HALF-LIFE of 2.1 million years—could sustain a chain reaction (compared to a "critical mass" of 10 kilograms for plutonium-239 and 50 kilograms for URANIUM-235). Extracting enough for a weapon would be even more arduous than working with those two old warhorses, however, so it has never been done and probably never will be.

In 1999 the International Atomic Energy Agency—an arm of the United Nations that audits and inspects nuclear materials around the world—called for voluntary reporting of inventories of neptunium and AMERICIUM that have been extracted from spent nuclear fuels, in order to strengthen safeguards against clandestine weapons development. Commercial nuclear reactors generate tons of neptunium every year in waste, and some 340 kilograms of pure neptunium are a legacy of the American nuclear weapons program. Presumably, the Soviet Union bequeathed a similar amount.

Neptunium plays a spoiler role in the unsolved problem of how to dispose safely of highly radioactive waste. Heat generated in the first 1,500 years of storage limits how much can be packed into any repository. The decay of plutonium-241 to americium-241, which then decays to neptunium-237, is the main source of heat during the first thousand or so years. Too much of this material will heat up the rock of a sealed underground dump to destructive levels. As for long-term radiation hazard, the same plutonium-to-americium-to-neptunium decay chain leaves the longest-lived element, which would need to be isolated from the biosphere for hundreds of thousands of years. So far, no one knows where to put this deadly stuff, which continues to pile up at reactor sites worldwide.

The americium-241 in common household smoke detectors will eventually turn into neptunium-237, but this is another problem being passed on to future generations.

Neutrons

ABORIGINAL INHABITANTS OF THE ATOMIC nucleus, discovered in 1932 by the English physicist James Chadwick (1891–1974). Within six years, doctors were trying to zap cancer cells with neutrons, but this was not destined to be neutrons' most momentous application. Because the particles have no electric charge and can therefore smash effortlessly into atoms and split nuclei apart, they led directly to the discovery of FISSION in 1939. By the early 1940s, with World War II raging across Europe, a few elite scientists understood the feasibility of atomic bombs.

In a 1969 interview, Sir James recalled the emotional effect this nightmarish revelation had on him. It was a rare public admission by any of the prominent scientists who helped create the Bomb that their work took a psychological toll:

> I remember the spring of 1941 to this day. I realized then that a nuclear bomb was not only possible—it was inevitable. Sooner or later these ideas could not be peculiar to us. Everybody would think about them

before long, and some country would put them into action. And I had nobody to talk to. You see, the chief people in the laboratory were [Otto] Frisch and [Jozef] Rotblat. However high my opinion of them was, they were not citizens of this country, and the others were quite young boys. And there was nobody to talk to about it; I had many sleepless nights. But I did realize how very, very serious it could be. And I had then to start taking sleeping pills. It was the only remedy; I've never stopped since then. It's 28 years, and I don't think I've missed a single night in all those 28 years.

The initial radiation from an atomic bomb consists of a small flood of neutrons plus a tidal wave of GAMMA RAYS. Neutrons were more significant at Hiroshima than at Nagasaki, due to the use of uranium instead of plutonium fuel. Today, exposure to neutrons comes mostly from the fission reactions in nuclear power plants, from COSMIC RAYS, and from medical radiotherapy or industrial radiography (inspecting, testing, or viewing inside materials without destroying them). Medical use of neutrons is relatively limited compared to X-RAYS.

The fact that neutrons are electrically neutral means that they are deeply penetrating. They do not damage cells in living tissue by ionizing molecules themselves, but by bashing into the nuclei of hydrogen atoms, which fly away like billiard balls and wreak extensive havoc. Neutrons are therefore weighted more heavily than X-rays or gamma rays in the calculation of EFFECTIVE DOSE. That is, they are more carcinogenic. Their exact rating—5, 10, or 20 (the same as ALPHA PARTICLES)—depends on how energetic they are. Exposure to high-energy neutrons is therefore different from exposure to low-energy ones. For example, cosmic rays carry high-energy neutrons, but nuclear reactors produce the low-energy kind. All are considered high-LET radiation—that is, very dangerous.

Individual lifetime (in this case, seventy-five years) neutron doses have been estimated for various natural and man-made sources. For cosmic rays, the worldwide average is about 6 mSv, but residents of high-altitude cities may collect as much as a quarter of their total exposure to natural background radiation from neutrons. La Paz, Bolivia, holds the record, at 68 mSv. Professional couriers who spend a lot of time on long-haul airline flights can rack up as much as 46 mSv. Workers in the civilian and military nuclear industry average about 44 mSv, with maximums around 130 mSv.

On those occasional dreary days when life seems too hard, we may all thank our lucky stars that the earth's atmosphere is such a good shield against cosmic rays—usually, that is. During the largest solar flare event yet observed, on February, 23, 1956, the neutron count at ground level rose 3,600 percent above normal background rates.

Nickel-59, -63

ONLY TWO RADIOISOTOPES of the element nickel have a long enough HALF-LIFE to be worrisome: nickel-59 at seventy-five thousand years and nickel-63 at one hundred years. Both are in the radioactive waste that comes from reprocessing spent nuclear reactor fuel. When NEUTRONS from FISSION reactions in the core strike various reactor components made of steel alloys that contain elements such as chromium, manganese, nickel, iron, or cobalt, such as the fuel rod structures themselves, they may be absorbed and spawn many different radioactive substances, including nickel-59 and -63. Nickel-63, a BETA PARTICLE emitter, is produced in higher concentrations. Trace amounts are also found everywhere from FALLOUT.

Niobium-95

NIOBE, DAUGHTER OF TANTALUS, from Greek mythology, is the namesake of the element niobium, which is chemically similar to the element tantalum and was identified in a sample of tantalite ore in 1846. Known earlier in that century as columbium, niobium was not officially adopted by the International Union of Pure and Applied Chemistry until 1950. The RADIOISOTOPE niobium-95—a BETA PARTICLE emitter with a HALF-LIFE of thirty-five days—is an important FISSION product in the waste effluent discharged from power plants that reprocess spent nuclear reactor fuel. It is often found together with ZIRCONIUM-95—zirconium alloy is commonly used in the cladding of reactor fuel rods—from which it forms as a DECAY PRODUCT. Because zirconium-95 has a longer half-life (sixty-four days), niobium-95 in reactor waste disappears more slowly than its own half-life would account for. Niobium-95 can also be released from damaged reactors and was among the radioactive contaminants detected in soil after the Fukushima Daiichi disaster.

Occupational radiation

I T IS REASONABLE TO ASSUME that most people avoid radiation if they can. If it's part of their job, however, being unexposed might mean being unemployed. Workers need the protection of regulatory laws to prevent not just injury but exploitation. Since 1928 the independent International Commission on Radiological Protection (ICRP) has issued recommendations on exposure limits that are widely adopted by governments. Of course, laws can be evanescent. Soon after the March 2011 Fukushima nuclear reactor disaster, the Japanese government abruptly raised the standard exposure limit for nuclear workers by 150 percent, from 100 mSv (averaged over five years) to 250 mSv. (The limit for the general public, including children, was also raised, from 1 mSv to 20 mSv.) Even so, by the end of May, at least eight emergency workers had been exposed to more than 250 mSv of internal (inhaled or ingested) radiation, ranging as high as 580 mSv for some individuals. Ninety-four others exceeded the original legal limit with combined internal and external doses. TEPCO, the utility company that operated the plant, admitted in June that it could not even find sixty-nine of the workers for radiation checks and that the names of thirty of these did not appear in company records. Transient day laborers, many with links to "yakuza" crime organizations, were performing the most dangerous—that is, highly irradiated—jobs.

While nuclear catastrophes such as Fukushima cause the most acute occupational exposures, millions of workers around the world are exposed during the course of their normal activities. In general, they may encounter higher-than-usual levels of natural BACKGROUND RADIATION, as happens especially in underground mining, or risky man-made sources, as in the nuclear power, weapons, and medical industries. RADON-222 is the biggest such peril in most mining operations. Indeed, numerous long-term epidemiological studies of miners around the world form the basis for estimating health risks from breathing radon and show clear associations between exposure and lung cancer. But various other RADIONUCLIDES with a long HALF-LIFE found in minerals not commonly thought of as being radioactive at all can create surprising hazards along the path from raw material to end product.

There are thousands of radioactive isotopes, but only a dozen or so are of concern in man-made devices: the so-called "University Five" (CARBON-14, POTASSIUM-32, IODINE-125, IODINE-131, and CALIFORNIUM-252), which are found in biochemistry labs and medical practice; the "Industrial Three" (IRIDIUM-192, CESIUM-137, and COBALT-60), which are used in radiography and other processes where penetrating GAMMA RAYS are useful; and the "Military Five"

(TRITIUM, URANIUM-235 and -238, PLUTONIUM-239, and AMERICIUM-241), which are found in the military-industrial complex.

The standard whole-body exposure limit promulgated by the ICRP is 20 mSv per year averaged over five years, and 50 mSv in any single year. Doses to the lens of the eye should not exceed 150 mSv, to the skin 500 mSv, and to the hands or feet 500 mSv. These occupational limits are much higher than what is recommended for the general public (1 mSv per year), which includes more vulnerable individuals such as children and pregnant women.

Most, if not all, of the industries where radiation is of inherent concern would grind to a halt if exposures were eliminated. Airline crews are one of the most highly exposed occupational groups, due to COSMIC RAYS at high altitudes, but obviously they will not stop flying. Mining is clearly a dangerous job no matter what, even with radon not at the top of the hazard list. The problem thus tends to be overlooked. Close and systematic monitoring of miners is hardly the norm in an industry notorious for skirting safety regulations. Publicly available data about exposures are usually expressed in annual averages, which mask the high end. For example, the estimated yearly average exposure of 123,333 gold mine workers in South Africa was 7 mSv in 2000, but 3,700 of these men received doses greater than 20 mSv. Diamond miners in that country are routinely X-rayed to see if they are stealing gemstones by hiding them inside their bodies, but there are no reliable numbers for how many men or how often. Sometimes the averages alone are alarming: phosphate miners at some locations in Egypt average between 107 and 182 mSv per year from inhaled radon.

Oil and gas workers, too, are exposed to natural radionuclides in the gushers that rise to the surface from the earth's crust. RADIUM-226 and -228 and their associated DECAY PRODUCTS are of chief concern. Dose rates a thousand times higher than normal background values have been found around production sites. Welding is an occupation not commonly associated with radiation exposure, but the widespread use of tungsten welding rods containing thorium can result in many times the level of natural background radiation. Phosphate fertilizers are another common source perhaps not ordinarily connected with radiation in the public's mind. White porcelain tiles containing zircon may radiate as much as 120 µSv per year. And for those who dream of Indiana Jones adventures, it might be worth considering that archaeological workers in Egyptian pyramids and tombs may receive annual doses as high as 13 mSv from radon. All in all, about thirteen million workers around the world encounter higher-than-normal natural sources of radiation, sometimes *much* higher.

Compared with mining, the monitoring of workers in the nuclear power industry is more comprehensive. Various international organizations maintain detailed databases. Vulnerable tasks run the gamut from uranium mining (similar to any other type) and enrichment, to fuel fabrication, to reactor operation and maintenance. Waste disposal, the great Achilles' heel of nuclear power, is still so undeveloped that one can only guess its effects on workers, let alone the public. About nine-tenths of the world's annual production of uranium comes from just ten countries (Australia, Canada, Kazakhstan, Namibia, Niger, Russia, South Africa, Ukraine, United States, and Uzbekistan), led by Canada, at 30 percent. Average annual exposures tend to be in the low single-digit milliSIEVERT range from radon or ALPHA PARTICLE–emitting dusts. Enrichment is a more closely held process dominated by just five suppliers in the United States, France, Russia, China, and a consortium of Germany, United Kingdom, and the Netherlands. Average annual exposures are generally in the low tenths of a millisievert. Numbers for fuel fabrication depend on whether officials record internal exposures from inhaled materials or external exposures to GAMMA RAYS, or both, but what figures exist are in the low single digits.

As of January 2011 there were 442 nuclear power plants operating in 30 countries, with 65 more under construction. How much the Fukushima debacle will change this picture remains to be seen. (The German government announced in May 2011 that it would phase out all its nuclear power plants by 2022; in Japan, former prime minister Naoto Kan said in July 2011 that the country should wean itself completely, with polls showing that more than two-thirds of Japanese citizens support this goal.) These reactors vary in age, design, and output, but they are all such exceedingly complex devices that building and controlling them is fraught with error, as history has shown. External exposure to gamma rays from COBALT-58 and -60 and SILVER-110 is the primary problem for workers, especially during maintenance or refueling operations. Serious accidents that cause cooling system breakdowns, such as at Three Mile Island and Fukushima, may release core FISSION products such as CESIUM-137 and ZIRCONIUM-95. In some types of reactors, circulating coolant water carries NITROGEN-16. Tritium is produced in others that may cause internal exposures to BETA PARTICLES. In 1997, tritium concentrations thirty-two times higher than federal drinking water standards were discovered in wells near a research reactor at the Brookhaven National Laboratory on Long Island, New York.

The most common reactor design around the world is the 1950s-vintage light-water moderated-and-cooled type (abbreviated PWR or BWR, depending on whether the circulating coolant water is pressurized or allowed to boil for steam). The name refers to the coolant system and how neutrons from the reactor core are slowed down to create useful heat for turning electrical dynamos. Developed by General Electric, most boiling water reactors are in the United States and Japan—the damaged Fukushima reactors were all first-generation BWRs built in the 1970s. External doses of gamma radiation from fission products are the primary concern for the more than one hundred thousand workers at these plants, with most exposures taking place during planned and unplanned maintenance shutdowns.

The annual exposure averages associated with working at commercial nuclear reactors are nowadays in the low single-digit millisievert range, having decreased significantly from double-digit figures in the 1970s thanks to stringent regulations and technical improvements. Of course, the industry's history shows that when things go wrong, they go very wrong, and workers are always highly vulnerable.

There are many other industries that use radiation in which occupational exposures may or may not be systematically monitored and reported. For example, irradiating medical products to sterilize them, food to preserve it, and even insects to exterminate them, is common around the world, but data about these practices are scarce. This is especially worrisome, because these industries entail very high quantities of radiation from cobalt-60 or cesium-137 that must be carefully controlled to protect workers. Radiography, where gamma rays or X-RAYS are used to find defects in pipelines and machinery, is ubiquitous and subject to all the vagaries of workplace safety. Luminization—creating "glow-in-the-dark" numbers and letters by mixing alpha or beta emitters with a phosphor such as zinc sulfide—is one of the oldest uses of radiation and the cause of the infamous radium poisoning of women who painted watch dials at a New Jersey factory in 1917. (Litigation eventually established landmark labor safety standards.) Perhaps not surprisingly, most modern luminization data come from Switzerland, showing that while average exposures have declined in the last generation

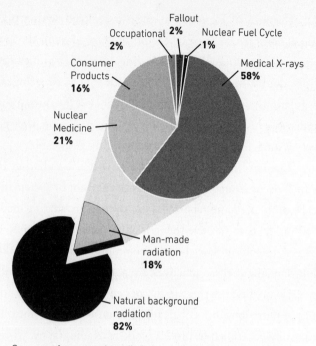

Sources of man-made radiation

from double-digit to single-digit millisieverts, doses over 15 mSv are not uncommon.

Finally, there are the military workers, in uniform or not, who make or handle atomic weapons, operate the nuclear reactors that propel ships, and do many of the same jobs with radiation that are found in civilian industries. Exposure data for these hundreds of thousands of people are far from comprehensive, dominated by information from the United States (which has not built new nuclear weapons since 1992) and the United Kingdom, showing that occupational exposure has fallen over the years to low single-digit millisieverts and less. But given the way secrecy and bureaucracy are used to cover error and embarrassment in the military, no observer should take these numbers on faith.

A healthy dose of skepticism, in fact, should accompany any information about occupational radiation exposure.

Plutonium-238, -239, -240, -241, -242, -244

ONE BREUGELIAN VERSION OF HELL is the process of recovering plutonium from the spent uranium fuel rods of nuclear reactors. From 1944 to 1988 the U.S. government produced more than 100,000 kilograms of radiotoxic plutonium-239 to make nuclear weapons. First they dissolved the fuel rods—containing plutonium at about 250 parts per million—with nitric acid, then passed the slurry through a noxious series of chemical reactions to separate the plutonium from uranium and other FISSION products. All work was performed by remote control, in specialized buildings so vast they were called "Queen Marys." The job produced Augean quantities of chemical and radioactive waste: about 10,000 gallons per 1,000 kilograms of spent fuel, for a total of some 100 million gallons of exceedingly hazardous muck that will remain part of the societal burden of the cold war arms race for tens of thousands of years. The Soviet Union generated its own such nightmare, of course, on a similar scale.

Plutonium is essentially the wicked witch of the nuclear

era, a nasty gift of the twentieth century to the planet's fore-seeable future. It has some of the weirdest properties of any element, such as shrinking as it gets hotter and conducting less heat or electricity than any other metal. A few primordial isotopes (such as plutonium-244, whose HALF-LIFE is 80.8 million years) exist in vanishingly small traces, and there are others that can in theory be produced naturally on rare occasions, but they may be regarded as insignificant. Any plutonium now in the biosphere is a man-made poison. The collective genius of the Manhattan Project bequeathed the knowledge that 1 kilogram of plutonium-239 can produce an explosion whose energy is equivalent to about 15 million kilograms of TNT, and life has never been the same since. About 6.2 kilograms were used in the "Fat Man" bomb dropped on Nagasaki, Japan, in August 1945.

If there were one element that could be yanked from the periodic table forever, plutonium would surely be it. Its few humane uses, such as plutonium-238 in the small thermo-electric generators of interplanetary space probes, are trivial compared with its military applications and attendant environmental pollution. Glenn Seaborg and his team of radio-isotope sleuths fathered it in 1940—they called it plutonium because it follows neptunium in the periodic table. Designed to be dangerous, it was then manufactured under secret federal auspices for the first atomic bomb test ("Trinity") in July 1945 and the attack on Nagasaki the following month. FALLOUT from subsequent decades of atmospheric weapons tests scattered about 10,000 kilograms of various plutonium isotopes worldwide, even into the bodies of everyone alive since 1945. Wherever plutonium is found today, it is a legacy either of nuclear weapons or of accidents at nuclear reactors.

Within weeks after the disaster at Japan's Fukushima Daiichi reactor plant, plutonium-239 was found in nearby soil. From a hundred soil samples taken within an eighty-kilometer-radius zone between June and July 2011, plutonium-238 turned up in six locations, at concentrations ranging from 0.55 to 4.0 BECQUERELS per square meter. Soil from the evacuated village of Iitate, about sixty kilometers from the damaged reactors, held plutonium-238 at 0.82 Bq/m², and both plutonium-239 and plutonium-240 at 2.5 Bq/m². Most of it probably came from the meltdown of Reactor 3, whose core contained so-called MOX (for "mixed oxide") fuel composed of a mélange of plutonium isotopes (about 8 percent 239 and the rest 240, 241, 242, which are considered useless contaminants). All are potent emitters of ALPHA PARTICLES, except 241, which gives off BETA radiation. Burning MOX in commercial nuclear power reactors has been touted for disposal of weapons-grade plutonium from cold war stockpiles, but the practice is still limited. (Ordinary low-enriched uranium fuel contains primarily URANIUM-238, the most common natural isotope of uranium, along with about 5 percent URANIUM-235, which fissions more easily. MOX fuel incorporates plutonium-239 as the fissionable component, lowering the need for uranium-235.) Based on an agreement with Russia in 2000 for each nation to dispose of surplus weapons-grade plutonium, the

U.S. Department of Energy oversees a $5 billion project to recycle 34,000 kilograms of plutonium-239 into MOX, but the fuel has not yet been used in any American power plants. Critics fear that burning fuel enhanced with plutonium raises the risk to the environment in the case of an accident, a point that appears to have been borne out at Fukushima.

Plutonium materials are classified into four grades according to how much undesirable plutonium-240 they contain: reactor grade (greater than 19 percent), fuel grade (7 to 19 percent), weapons grade (less than 7 percent), and super grade (2 to 3 percent). In a conventional nuclear reactor, a kilogram of plutonium-239 produces enough heat to generate almost 10 million kilowatt-hours of electricity; thus, a typical 1,000 megawatt commercial plant contains within its uranium fuel rods several hundred kilograms. Fuel- and reactor-grade plutonium can still be used to make nuclear weapons, though of inferior type. In 1994, Secretary of Energy Hazel O'Leary revealed that the United States had successfully tested a bomb made with reactor- or perhaps low fuel-grade plutonium from Great Britain in 1962. All grades could be used in so-called dirty bombs, which are made with conventional explosives to contaminate wide areas. About 80 percent of the 100,000 kilograms of plutonium that remain in American stockpiles is weapons-grade, primarily plutonium-239.

From 1945 to 1947, without informed consent, eighteen people who were expected to live fewer than ten years were injected with between 5 and 100 micrograms of plutonium at the Manhattan Engineer District Hospital in Oak Ridge, Tennessee (one patient), at Strong Memorial Hospital in Rochester, New York (eleven patients), at Billings Hospital of the University of Chicago (three patients), and at the University Hospital of the University of California in San Francisco (three patients). Excreta were obtained from these human guinea pigs and sent to Los Alamos for plutonium analysis. The results were used to establish mathematical equations describing plutonium excretion rates. The government kept the identity of the subjects secret for decades— five were identified in news articles in 1993—out of concern for public relations and legal liability. Seven lived longer than ten years, and five longer than twenty years. Survivors were not told until 1974 that they had been injected with plutonium.*

* See *American Nuclear Guinea Pigs: Three Decades of Radiation Experiments on U.S. Citizens*, U.S. House of Representatives, Committee on Energy and Commerce, November 1986, and W. H. Langham, H. Bassett, P. S. Harris, and R. E. Carter, *Distribution and Excretion of Plutonium Administered Intravenously to Man*, Los Alamos Scientific Laboratory, LA1151, republished in *Health Physics* 38 (1980).

Polonium-210, -214, -218

A FEROCIOUS EMITTER OF ALPHA PARTICLES that is regarded as one of the most dangerous substances known, polonium was the first radioactive element discovered by Marie and Pierre Curie, in July of 1898, followed by RADIUM in December. It epitomizes all the double-edged swords of nuclear phenomena and might be the most notorious radioisotope if URANIUM and PLUTONIUM bomb-making had not sunk the entire field into devilish territory.

The inimitable Curies were trying to understand why unrefined pitchblende ore was more radioactive than the uranium that they could extract from it. After an intricate sequence of chemical separations, they obtained the culprit—four hundred times more radioactive than uranium—and named it after Marie Sklodowska Curie's native Poland, which was then partitioned under foreign rule. The lab work was not exactly delicate: a ton of uranium ore contains only about 100 micrograms of polonium. In a decrepit outdoor shack that one visitor likened to a potato shed, they labored more like Bohemian miners than elite scientists, with Mme. Curie processing more than forty pounds of rock at a time. "Sometimes I had to spend a whole day stirring a boiling mass with a heavy iron rod nearly as big as myself," she wrote. "I would be broken with fatigue at day's end." Perhaps this is why they made the discovery instead of their social and professional superior, Professor Henri BECQUEREL, on whom they nonetheless depended to publish their momentous findings in the journal of the French Academy of Sciences, *Comptes rendus*. Mme. Curie's use of the word *radio-active* in the paper's title—"Sur une nouvelle substance fortement radio-active, continue dans la pechblende"—was the first time the term appeared in any language. Both she and Pierre suffered debilitating health effects from their toil with radioactive materials, unaware of the terrible risk they were taking.

Polonium has more than thirty isotopes, all radioactive. Polonium-210 occurs naturally, with a HALF-LIFE of 138 days, as a DECAY PRODUCT of ubiquitous uranium-238, along with two other polonium radionuclides, 214 and 218, which have far quicker half-lives (3 minutes and 164 microseconds, respectively). Though rare, polonium-210 is the dominant member of the family—214 and 218 would be of merely academic interest if not for the fact that they are RADON "daughters" that can lodge in lung tissue and deliver highly carcinogenic doses of alpha particles. For commercial purposes, nobody today refines uranium ore the way Marie Curie did to obtain polonium. Nuclear reactors are employed instead, to transmute the element bismuth by bombarding it with NEUTRONS.

Polonium-210 is extremely "hot" in both the radioactive and thermal sense. One milligram emits as many alpha particles as 5 grams of RADIUM-226, though almost no GAMMA RAYS. This made it attractive to Manhattan Project bomb designers, who incorporated it into the "initiator" device that produced a precisely timed burst of neutrons to start an explosive chain reaction. (Beryllium absorbed the alpha particles and then spewed neutrons into the bomb's ENRICHED URANIUM core.) In a humane rather than destructive application, the heat energy from half a gram could stoke its container to 500°C and provide useful energy. Moreover, a pound formed into a three-cubic-inch chunk to fuel a thermoelectric generator could turn out 64 kilowatts of electricity. These possibilities are limited in practice by polonium-210's short half-life, but have found utility in so-called SNAP (Systems for Nuclear Auxiliary Power) generators that power space satellites and probes.

The presence of polonium-210 in tobacco leaves, and thus cigarette smoke, casts one of the most treacherous shadows across an industry that actively suppressed knowledge of smoking's dangers for decades. In 1964 a research paper published in the journal *Science* reported the existence of polonium-210 in tobacco smoke. Smoking one and a half packs a day for thirty years would expose the lungs to as much radiation as three hundred chest X-RAYS a year, causing an estimated 1 percent of all lung cancers in the United States—or about 2,225 new cases in 2010. (Much of the excess cancer attributed to breathing RADON may in fact be due to polonium-210—smokers are killed by alpha radiation, wherever it comes from.) The high-phosphate fertilizers that are used to grow tobacco contain radium-226, one of whose decay products is polonium-210, which is absorbed by plant roots and via fertilizer dust on leaves. The major tobacco companies tried for decades to remove or filter polonium-210, but gave it up as commercially infeasible. According to a 2008 analysis in the *American Journal of Public Health* of internal corporate documents released in response to litigation, the companies followed a legal strategy whereby avoiding knowledge of biologically significant levels of polonium-210 in their products would allow them to ignore it as a possible cause of lung cancer. They thus stymied efforts by their own scientists to perform or publish further research on the subject, seeking to "avoid heightening the public's awareness of radioactivity in cigarettes." Ignorance was bliss.

It is rather ironic that this magnificent radioisotope has found industrial applications of utmost mundanity. For example, because the alpha particles emitted by polonium will ionize surrounding air, brush hairs containing minute quantities are handy for sweeping staticky dust off films and lenses. The static electricity generated by large rolling sheets of paper or cloth in a press or mill can also be neutralized by polonium sources. Perhaps the most exotic use of polonium-210 was as an assassination weapon, when Russian dissident and former KGB officer Alexander Litvinenko was murdered by Russian state agents in 2006. Polonium-210 is highly toxic, extremely radioactive (sixty-five thousand times more than PLUTONIUM-239), and easy to ingest. While at the Millennium Hotel in London, Litvinenko apparently drank tea that had been spiked with about 10 micrograms of polonium-210, which killed him three weeks later. The case is still officially open in London, due to the refusal of the Russian government to extradite the prime suspect, ex-KGB agent Andrei Lugovoi. "We still disagree with you over the Litvinenko case," British prime minister David Cameron said diplomatically in a speech at Moscow State University in September 2011.

Potassium-40

THIS IS ONE RADIOISOTOPE there is no getting away from. You wouldn't want to, anyway. We need the element potassium for healthy metabolism, and this radioactive isotope comprises about 0.012 percent of it. It is the main source of natural radioactivity in the human body and in most food. A typical 70-kilogram adult contains about 140 grams of potassium, of which 0.017 gram consists of potassium-40. This hot little load of atoms disintegrates at the rate of about 266,000 per minute. Of every one hundred atoms that disintegrate, eighty-nine release BETA PARTICLES and eleven give off GAMMA RAYS, which means the annual EFFECTIVE DOSE amounts to about 165 microsieverts. Children, with a bit more potassium in their bodies, get about 185 μSv. Potassium-40's HALF-LIFE is 1.3 billion years, so it isn't going anywhere soon.

Though the amount of potassium-40 in soil varies geographically (15–990 BECQUERELS per kilogram in Japan, for example, and 100–700 Bq/kg in the United States) and gets boosted wherever farmers apply common fertilizers, the amount of potassium in the body is regulated homeostatically to be constant. Eating foods that are comparatively high in potassium-40—such as cantaloupe, Brazil nuts, bananas, apricots, or acorn squash—may increase internal exposure for a few hours, but soon the kidneys restore the normal level through excretion. Long-term exposure is therefore stable. A healthy adult male carries a steady level of about 0.1 micro-CURIE of potassium-40.

Praseodymium-144

WHEREVER THERE IS CERIUM-144, there you will find its daughter with the odd-looking name (pronounced PRAY-zee-oh-DIM-ium). The element praseodymium was discovered in 1885 by the Austrian aristocrat chemist Baron Carl Auer von Welsbach (1858–1929), who separated it from a mineral called didymium—*prasios* means "green" in Greek, referring to the color of its salts. The HALF-LIFE of praseodymium-144, a BETA PARTICLE emitter, is only 17.28 minutes, but cerium-144's longer existence (half-life: 285 days) keeps it around.

The baron was perhaps more interesting than his discovery. In the 1890s he invented and commercialized an incandescent mantle for gas lamps using thorium and cerium oxides obtained from monazite sand (a source of high levels of natural BACKGROUND RADIATION, unknown at the time, due to its URANIUM content). German freighters leaving Brazil loaded the sand—whose value as a source of thorium and cerium was unappreciated—as ballast, giving the wily von Welsbach a free supply. He later developed metallic filaments for electric lights, so-called flints for gas lighters (actually made of cerium alloys), and large-scale chemical separations of radioactive elements. After his death, the company he founded produced Doramad, a thorium-laced radioactive toothpaste, advertised as follows:

> Its radioactivity increases the defenses of teeth and gums. The cells are loaded with new life energy, the bacteria are hindered in their destroying effect. This explains the excellent prophylaxis and healing process with gingival diseases. It gently polishes the dental enamel so it turns white and shiny. Prevents dental calculus. Wonderful lather and a new, pleasant, mild and refreshing taste. Can be applied sparingly.

Protactinium-231

A NATURAL RADIOISOTOPE identified in 1917–18 by Lise Meitner and Otto Hahn, who twenty years later would play seminal roles in discovering nuclear FISSION. The element's name comes from the Latin *protoactinium*, meaning "before actinium," pointing to the fact that ACTINIUM-227 is a DECAY PRODUCT of protactinium-231. Itself a decay product of URANIUM-235 via THORIUM-231, protactinium-231—an ALPHA PARTICLE emitter with a HALF-LIFE of thirty-three thousand years—contaminates sites where uranium ore has been processed, and spent nuclear reactor fuel. Expensive to produce in pure form and highly radiotoxic, it has no practical uses.

Radioactive waste

THERE IS NO PERMANENT DUMP for old nuclear fuel. More than the threat of disastrous reactor leaks, which after all have been unusual, this is the single most critical Achilles' heel for the nuclear power industry. After more than half a century, there is still nowhere to put the highly radioactive waste produced by nuclear reactors so that it remains safely isolated from the human environment for the thousands or even millions of years that it will be dangerous. About 250,000 tons of the hot stuff is now stacked underwater on-site at reactor centers around the world, with all the attendant risks that were starkly evident during the Fukushima debacle, when storage pools leaked catastrophically.

Of course, guaranteeing the removal of this material from the biosphere for millennia is a technical and political absurdity. Like interstellar space travel, it awaits a miraculous solution. Meanwhile, waste will keep piling up, at about 10,000 tons a year, as long as just the current number of nuclear power plants exist. And then it will stay around, practically speaking, forever.

For short-term management purposes, radioactive waste

is divided into several categories. Very-low-level waste (VLLW)—common building materials from dismantled or refurbished nuclear industrial sites—is not considered harmful and is handled like any other junk. Low-level waste (LLW) consists of hospital and industry refuse: rags, paper, clothing, tools that contain short-lived radioactivity and do not require shielding. LLW can be safely buried in shallow earth pits. Intermediate-level waste (ILW) is radioactive enough to need special shielding such as concrete: chemical sludge, decommissioned reactor parts. And high-level waste (HLW) is the nastiest: burned-out fuel from reactor cores that contains FISSION products that are strong emitters of BETA PARTICLES and GAMMA RAYS. It is thermally as well as radioactively "hot," thus requiring continuous cooling by circulated water. If this fails and the old fuel is exposed to the air, it will soon combust and release prodigious quantities of CESIUM-127.

A typical 1,000-megawatt reactor generates about 27 tons of spent fuel a year. Some of it may be reprocessed to make new fuel, but most goes into the local storage pool, which is always right next to the reactor, without its own CONTAINMENT structure. Some 1,760 tons were on-site at Fukushima, with no backup water-circulation systems. At the Vermont Yankee power station in the United States, which is of the same design, some 690 tons of spent fuel are on-site today.

A somewhat safer alternative to pool storage for HLW is known as "dry cask" storage, in which used fuel rods are sealed in steel barrels along with inert gas and then placed in concrete containers. Germany has used this method for twenty-five years, but its higher cost has been a barrier elsewhere. It is by no means a perfect solution. During a 5.8-magnitude earthquake in Virginia in August 2011, 113-ton dry casks at the North Anna nuclear power plant were knocked four inches out of place—more than enough to raise concerns about their structural integrity.

The waste from reactors may be the worst problem, but it is not the only one. According to the U.S. Department of Energy, "fifty years of nuclear weapons production and energy research generated millions of gallons of radioactive waste, thousands of tons of spent nuclear fuel and special nuclear material, along with huge quantities of contaminated soil and water." Hundreds of sites across the country will remain expensive technological and political problems for many generations to come.

Even if technical solutions can be found, political considerations may derail them. Such has been the case with the proposed Yucca Mountain deep-underground HLW repository in Nevada, which was approved by Congress in 2002 after twenty-four years of study and canceled in 2009 due to intense state opposition. Not in my backyard.

Radioisotope, radionuclide

WO WORDS FOR ESSENTIALLY the same thing. Why? Perhaps because physicists and chemists get separate Nobel Prizes. All that really matters is that atoms of the same element sometimes hold different numbers of NEU-TRONS. In 1913 the British chemist Frederick Soddy (1877–1956) named these close family members isotopes—from the Greek words meaning "equal place"—because they occupy the same square in the periodic table. Working with Ernest Rutherford, Soddy showed that radioactivity entails the "transmutation" (a term as ancient as the alchemistic quest for a magical "philosopher's stone" for turning common metals into gold), or conversion, of elements into DECAY PRODUCTS, which may be an isotope of a different element or a different isotope of the same element. In 1921 he won the Nobel Prize in chemistry for this research. Rutherford got his Nobel, also in chemistry, in 1908, though he is revered as a great physicist.

"Put colloquially, their atoms have identical outsides but different insides," Soddy explained about isotopes in his Nobel lecture, showing typical willingness to speak plainly.

Turning away from radiochemistry later in life to ponder the social implications of atomic energy, he believed that science was not value-free in the moral sense and that scientists ought not regard their profession as being ensconced in an ivory tower. With unusual ad hominem piquancy, his official Nobel biography describes him as "a man of strong principles and obstinate views, friendly with students and prickly with colleagues." Translation: his colleagues did not approve of his moonlighting in nonscientific subjects.

National Institutes of Health radiation safety poster

All isotopes are not radioisotopes—some are stable. If decay brings a change in the number of neutrons, the atom remains the same element, but becomes a different isotope of that element. All isotopes of one element hold the same number of protons but not the same number of neutrons. All isotopes of one element also have the same chemical properties but show different radiological characteristics such as HALF-LIFE or kind of radiation emitted when they decay. Radioisotopes are atoms of the same element that have unstable nuclei that decay into other elements, giving off various types of radiation in the process. Because they have the same number of electrons, they are chemically identical to one another, thus difficult to separate (see ENRICHED URANIUM). *Radionuclide* more often refers specifically to a nucleus and its unique physical characteristics than to a whole atom.

When electrons are added or removed from an atom, giving it a negative or positive charge, these forms of an element are called ions. This change alters the way the atom reacts chemically with other atoms, but not the stability of its nucleus (that is, its radioactivity).

There are about 3,700 radioisotopes/nuclides, both natural and man-made. They are symbolized by the element and its atomic weight (the number of protons and neutrons in its nucleus)—for example, plutonium-239, or Pu-239. To include the number of protons, called the atomic number, the symbol is $_{94}Pu^{239}$. Natural radioisotopes are either *primordial*, meaning that they are older than the earth itself, or *cosmogenic*, meaning that they originate from COSMIC RAYS. All elements with more than eighty-three protons in their nucleus are radioactive. Polonium—named as a political gesture in 1898 by Marie Curie after her native Poland at a time when it was not an independent nation—is the gateway to the radioactive realm of the periodic table (though technetium and promethium precede it), with eighty-four protons. It has thirty-three known radioisotopes.

Radioisotopes/nuclides are measured by their *activity*, rather than by mass, the way other hazardous materials are. The shorter a radioisotope's HALF-LIFE, the more radioactive it is. The standard unit of activity is the BECQUEREL.

The U.S. Environmental Protection Agency lists certain radionuclides as "commonly encountered": AMERICIUM-241, CESIUM-137, COBALT-60, IODINE-129 and -131, PLUTONIUM, RADIUM, RADON, STRONTIUM-90, TECHNETIUM-99, TRITIUM, THORIUM, URANIUM. Whether they are common or not, no one should ever be blasé about meeting them.

Radium-223, -224, -226, -228

RADIUM — THE GIVER OF RAYS — was the glory of Marie and Pierre Curie, who first separated the bewitching substance from URANIUM-laced pitchblende ore in 1898 at the dawn of the nuclear age. It was soon a death angel, too, as charlatans joined physicians in applying its mysterious powers to the human body. In 1900, Friedrich Giesel (1852–1927), who was praised as the "Patriarch of German Radium Research" by Nobel physicist Otto Hahn, published his childlike explorations in a scientific paper:

> I should like to confirm an observation . . . that radium rays, like Roentgen rays [X-RAYS], cause skin inflammations. I placed a double celluloid capsule containing 0.27 g of radium-barium bromide on the inner surface of my arm for 2 hours. At first there was only slight reddening; after 2–3 weeks, a strong pigmented inflammation appeared and, finally . . . the epidermis was sloughed off, whereupon healing soon followed.

Pierre Curie, who carried vials of purified radium—which glowed in the dark—in his vest pocket, duplicated the experiment himself. Soon doctors were using "radium burn" to kill skin tissue. Giesel died of radiation sickness, but radium-226 was still being applied to remove hemangiomas—generally harmless "strawberry" buildups of blood vessels that usually go away if left alone—in infants and children until the 1970s, a procedure since shown to have caused cancers of the thyroid, pancreas, endocrine glands, breast, and brain among these subjects. More

Pierre and Marie Curie (standing at left, much too close) with radium experiment, from a drawing by André Castaigne (1861–1929, who illustrated the first edition of *The Phantom of the Opera*).

than twenty-five thousand infants (younger than eighteen months) in Sweden were treated with radium for skin hemangiomas between 1920 and 1965.

In the United States, thousands of women were treated with intrauterine radium-226 between 1925 and 1965 for benign gynecological problems such as uterine bleeding, which resulted in excess deaths from leukemia and cancer of the colon, uterus, and genital organs. The ubiquity of radium in mainstream medicine is evident from a 1954 article by the chief of the U.S. Public Health Service's Radiological Health Branch, James Terrill, in *Public Health Report*:

> More than 80,000 radium treatments are given annually in the hospitals in the United States. A medium-sized modern hospital usually has about 300 millicuries (300 mg) of radium on hand for this purpose. Practicing physicians using radium will often have 25 to 50 millicuries, and nearly every dermatologist has 10 millicuries or more on hand.

The widespread use of radium in toothpaste, hair tonic, chocolate, suppositories, face creams, spring waters, phony elixirs, and sundry nostrums was a long, bizarre footnote in the history of twentieth-century radiotherapy. The mania was perhaps epitomized by the "radiendocrinator," which men were advised to place under the scrotum at night like an "athletic strap," for sexual virility. (Its inventor was William J. A. Bailey, a Harvard dropout who impersonated a doctor and claimed to have quaffed more radium water than any living person. He made a fortune from the sale of Radithor, a patent medicine of radium salts in water. He died of bladder cancer in 1949.) For decades, no application seems to have been considered too extreme by either quacks or MD's. An investigatory report by the U.S. Department of Energy in 1994 noted the following case involving institutionalized psychiatric patients:

> Patients in a state mental hospital were injected with radium as an experimental therapy for mental disorders. The experiment appears to have been conducted at the Elgin State Hospital in Elgin, Illinois, between 1931 and 1933. Documents indicate that 70 to 450 micrograms of radium-226 were injected. This experiment occurred prior to the establishment of the Argonne National Laboratory and the U.S. Atomic Energy Commission. Argonne National Laboratory later collected records and attempted to locate the subjects. Researchers believed that if the patients could be located and body content measurements made in the 1950s, a valid retention curve for radium in humans over several decades could be constructed. Argonne National Laboratory made all later measurements.

Argonne was apparently not interested in the subjects' health, but only in how much radium they still harbored.

Patient being "treated" with radium at London hospital, 1909

And then there was nasopharyngeal radium irradiation, or NRI, for chronic ear infections, hearing loss, sinusitis, tonsillitis, and asthma. From the late 1940s through early 1970s, as many as 2.5 million American children, according to Centers for Disease Control and Prevention estimations, had thin metal rods tipped with a platinum capsule of 50 mg of radium sulfate inserted through their nostrils to the rear of the nasopharynx near the opening of the Eustachian tube in order to irradiate and shrink adenoids and nearby lymphoid tissue. A typical course of treatment involved three insertions of about six to twelve minutes each, usually performed two to three weeks apart. During World War II and until the 1960s, some twenty thousand military submariners, divers, and aviators with ear problems were also treated. Between 1948 and 1954, in a federally funded experiment, Johns Hopkins Hospital administered radium in this fashion to 582 third-graders from the Baltimore City school system. The goal was to study the effect of radium on hearing loss. Radiation doses to adults (mostly from BETA PARTICLES and GAMMA RAYS, since the applicator capsule blocked ALPHA PARTICLES) have been estimated at 20,000–100,000 milliSIEVERTS to the mucosal lining of the nasopharynx, 240–1,200 mSv to the pituitary gland, 130–640 mSv to the salivary gland, 50–220 mSv to the brain, and 20–100 mSv to the thyroid. Doses to children would have been much higher.

Excess cancer risk was monitored in a long-term study by the Johns Hopkins School of Public Health that ended in 1979. The results caused the National Academy of Sciences to derive a cancer mortality risk factor of 8.8 excess brain cancer deaths over the lifetime of each 1,000 children treated with NRI.

The medical industry was not the only business mesmerized by radium. Until the 1960s it was the crucial ingredient of luminous paints used for clock dials, aircraft and military instrument panels, switches, and compasses—a ubiquitous commerce that left trails of environmental pollution and worker health scandals. Radium somehow became syn-onymous with progress and the power of science to deliver a better future, even as demonstrated by a cheap wristwatch.

All of radium's twenty-five isotopes are radioactive, originating as DECAY PRODUCTS of uranium or thorium. Radium-226 is a member of the uranium-238 decay series, while radium-228 and -224 are found among thorium-232's offspring. Radium-226, the most common one, emits alpha particles and gamma rays, with a HALF-LIFE of about 1,600 years. Radium-228 is principally a beta particle emitter and has a half-life of 5.76 years. Radium-224, another alpha source, dissipates by half in just 3.66 days. It has been linked to bone cancer. Adding to their injurious nature, isotopes of radium decay to form various forms of RADON gas: radium-226 to radon-222, and radium-228 via several steps to radium-224, before forming radon-220 (known as thoron).

Where is radium found today? Mostly in nature, where it belongs. Safer radioisotopes, such as COBALT-60, have re-placed it in medical practice, and it has been banished from consumer products.

Radon

COLORLESS, ODORLESS, TASTELESS — the quintessence of a threat beyond the reach of our senses. Radon is a radioactive gas found almost everywhere, accounting on average for half of all natural BACKGROUND RADIATION. It is the second leading cause of lung cancer after smoking, and the primary cause among nonsmokers. Tobacco smoke and radon are strongly synergistic carcinogens. Fortunately, just two of radon's three dozen isotopes exist in significant amounts in the environment: radon-222 and -220 (known as thoron). But this is quite enough to be the sharp focus of public health attention.

Radon is a DECAY PRODUCT of radium, which in turn is a decay product of uranium. Not until the post–World War II surge in uranium ore mining did medical literature turn away from happy talk about using radon therapeutically to concern about its harmful effects. In keeping with the general state of denial about radiation during the first half of the twentieth century, extensive historical evidence of radon's threat was ignored. As far back as the sixteenth century, as noted in *De re Metallica* (*On the Nature of Metals*) — a study

Colorless, odorless, tasteless, and where your children
play with their toys

of mining published posthumously in 1556 by the German
physician and town administrator Georg Bauer (1494–1555)
under the pen name Georgius Agricola—silver miners in
the Erz (literally "Ore") Mountains between Germany and
Czechoslovakia were known to be so plagued by fatal lung
disease that some women there had been widowed seven
times. Clearly a gifted observer, Agricola recommended ven-
tilating the mines with fresh air. But it took three more
centuries to identify the mysterious affliction as cancer, in
an 1897 article titled "Der Lungenkrebs, die Bergkrankheit
in den Schneeberger Gruben" ("Lung Cancer, Mountain
Sickness of the Schneeberg Miners"), by two German

researchers. Mortality rates meanwhile reached 50 percent in the "death shafts" there. Still, even as knowledge about radioactivity grew during the following decades and some doctors connected these cases with high concentrations of uranium in the ore, others felt free to blame genetic inbreeding among the insular miners.

The lucrative digging never stopped, of course. An Erz mine near Joachimsthal, Bohemia (now the Czech Republic), where Agricola had lived for nine years, supplied more than a ton of pitchblende (UO_2) to Marie Curie for her pioneering isolation of radium. Joachimsthal was also the site of the world's first "radon spa," in 1906, where vacationers partook of high-radon atmospheres for supposed invigoration. With the cold war accelerating in 1949 and the Pentagon frantic to obtain enriched uranium for nuclear weapons, the U.S. Public Health Service and the U.S. Atomic Energy Commission—aware of alarming lung cancer rates among European uranium miners—quietly launched a long-term study of miners in the Colorado Plateau, a region of major uranium deposits in Utah, Arizona, New Mexico, and Colorado. It took until the end of 1960, when the 1950s uranium boom was over and employment in the mines had already peaked, for the PHS to inform four western governors that the death rate from lung cancer among uranium miners was five times higher than that of American men in general. Other countries did not move so painfully slowly. France introduced forced-air ventilation of its uranium mines in 1956 and put occupational exposure regulations in place by

1958. Such standards were not implemented in the United States until 1971.

In 1966 the federal government did halt the practice in some areas of the United States of using uranium mill tailings rich in radium-226 to make concrete or as excavation backfill. In the 1970s, more than seven hundred homes and other structures around Grand Junction, Colorado, were razed to remove radioactive bricks or fill dirt. Under the Uranium Mill Tailings Radiation Control Act of 1978, the Environmental Protection Agency established an exposure threshold of 4 picoCURIES per liter of indoor air, above which the government would pay for cleanup of such houses.

Radon per se first gained national attention in the United States in 1984, when spectacularly high (650 times background) levels of radon were found inside a home near Boyertown, Pennsylvania, providing dramatic evidence that radon could seep out of the ground and into buildings, reaching concentrations similar to that at uranium mines. Concentrations were measured at 3,000 picocuries per liter (piC/L) of air—the equivalent in lung cancer risk to smoking 135 packs a day—after owner Stanley Watras set off radiation alarms at the Limerick nuclear power plant where he worked. By comparison, radon levels in Erz Mountain mines averaged about 2,900 picocuries and up.

In 1986, after estimating that about 10 percent of all the nation's homes probably held radon above 4 piC/L, the EPA issued advisories to the general public. This figure still stands as the yearly average at which the EPA recommends that homeowners do something (at their own expense) about indoor radon pollution. It is not a "safe" level, just a reasonably achievable level. The goal is to make levels inside as low as those outdoors, which vary geographically—averaging, for example, 5.4 in New Jersey, 0.8 in Texas, 0.12 in Washington, D.C. Average indoor exposure in the United States is probably 1 to 2 piC/L. Occupational limits are much higher, at 100 piC/L for miners and 30 piC/L for other workers. Even at 4 piC/L, however, large case-control epidemiologic studies have documented a 50 percent increased risk of lung cancer after prolonged exposure. The EPA estimates that if a thousand people who smoke were exposed to 4 piC/L for a lifetime (seventy years), sixty-two of them would develop lung cancer. For a thousand nonsmokers, seven would. At 2 piC/L, there would be thirty-two and four cancer cases, respectively.

In 2009 the Department of Energy finally began to remove by railcar some 16 million tons of radon- and gamma ray–emitting tailings dumped on the banks of the Colorado River by a uranium mill near Moab, Utah, that had been closed for twenty-five years. The transport to a site thirty miles away is expected to take until 2025 and cost at least $1 billion, though federal budget cuts have slashed the original workforce by two-thirds.

Technically, radon-222—the most important isotope because of its relatively long HALF-LIFE of almost four days— forms directly from ALPHA PARTICLE decay of RADIUM-226, which in turn comes from the URANIUM-238 that comprises

99.3 percent of all uranium in nature. The word *radon* derived from the term *radium emanation* during the early years of research on radioactivity by Friedrich Ernst Dorn (1848–1916), Ernest Rutherford, and other physicists in the first decade of the twentieth century. Both uraniun-238 and radium-226 are found in widely varying concentrations in most soils and rocks, making radon similarly ubiquitous. Certain types of granite, shale, gneiss, and phosphate rocks are the richest. In the United States, the Appalachians from Maine to Alabama (especially along the Reading Prong gneiss formations in Pennsylvania, New York, and New Jersey) are relatively high in radon, as are regions of granitic sediment in Iowa, California, Colorado, Idaho, and New Mexico. But high concentrations can appear anywhere.

Although chemically inert, radon is soluble in water and can thus also be found in drinking wells and especially in radium-rich springs such as those at Bad Kreuznach, Germany, and Misara, Japan, which continue to attract tourists lured by the charlatanism of "hormesis," the supposed beneficial effect of low-level, whole-body radiation. In Maine and Rhode Island, the average radon concentration in private well water exceeds 6,000 piC/L, contributing to indoor pollution through aeration in showers, toilets, and washers. Although there is a slight risk of developing internal organ cancers (mainly stomach) from drinking water with elevated levels of radon, the primary concern is release of the gas from normal water use into indoor air.

Research has shown that it is actually the short-lived

DECAY PRODUCTS of radon-222—sometimes called radon progeny or daughters, though this sort of gender imagery is fading—that are responsible for causing lung cancer. POLONIUM-214 and -218, LEAD-214, and BISMUTH-214 are solids that lodge in the lungs and bombard tissue with alpha particles. Thoron has such a short half-life that it does not have as much time to be inhaled, but one of its decay products, LEAD-212, exists long enough to spawn in turn the dangerous alpha emitter BISMUTH-212. (As if the usual assortment of exposure units were not confusing enough, measurements of radon in workplaces are sometimes expressed in "working levels" [WLs], one of which consists of any combination of radon progeny per liter of air that releases 1.3×10^5 million electron volts [MeV] of alpha energy. One WL = 100 pCi/L when the radon-222 and its progeny are in *equilibrium*, meaning the decay products undergo transformation at the same rate that they are produced.)

Easy-to-use radon gas test kits are widely available at reasonable cost.

Rubidium-82, -85, -87

THE ELEMENT RUBIDIUM, which is chemically similar to potassium and cesium, has one stable isotope
(rubidium-85) that comprises about 72 percent of all natural
rubidium. Its only other natural isotope, rubidium-87, makes
up the remaining 28 percent and is a primordial emitter of
BETA PARTICLES. Rubidium-87 has an enormously long
HALF-LIFE: forty-nine billion years, which is four times longer than the age of the universe, meaning that it is practically stable, too, though a blob of rubidium metal will fog
photographic film after a month or two. There are dozens of
artificial isotopes with short half-lives that are too radioactive for practical use outside the laboratory. Rubidium-82
chloride is sometimes used in nuclear medicine imaging of
the heart.

Ruthenium-103, -106

A METRIC TON of spent nuclear fuel contains more than 2 kilograms of the element ruthenium, a FISSION product of URANIUM-235. Ruthenium's longest-lived RADIOISOTOPES are ruthenium-106, whose HALF-LIFE is 374 days, and ruthenium-103, with a 39-day half-life—both are BETA PARTICLE emitters. For decades after the Soviet Union became a nuclear power, ruthenium was discharged in radioactive waste from the Mayak plutonium production plant into the Techa River in the southern Ural Mountains. More than twenty-five thousand residents of the region were exposed to this effluent, which caused increased cancer mortality from leukemia and risks for cancer of the esophagus, stomach, and lungs similar to those experienced by Hiroshima and Nagasaki survivors. The Russian chemist Karl Claus (1796–1864)—who discovered ruthenium in refining residues from Ural platinum ore in 1844 and named it after his homeland—might have altered the quaint practice of tasting his work if only he had known.

After the 1986 Chernobyl nuclear reactor explosion, highly radioactive ruthenium particles, more than 100 kiloBECQUERELS in activity and 10 μm in diameter, were found hundreds of kilometers from the plant. Ruthenium was also detected in water near the damaged Fukushima Daiichi reactor plant in 2011. Ruthenium-103 and -106 are among the RADIONUCLIDES monitored by the U.S. Food and Drug Administration in the food supply following a reactor accident. They were also released into the atmosphere by cold war–era nuclear weapons tests: 238 EBq and 11.8 EBq, respectively.

Because ruthenium is chemically similar to platinum, extracting it from nuclear waste might be commercially attractive if only it were not so radioactive. Лентяй. So it goes.

Scandium-46, -47

OBTAINED COMMERCIALLY AS A BY-PRODUCT of uranium refining, the element scandium consists of one stable isotope (scandium-45) in nature. It also has some two dozen artificial RADIOISOTOPES, of which only two possess half-lives of significant length: scandium-46, at 84 days, and scandium-47, at 3.35 days. Both are emitters of BETA PARTICLES and GAMMA RAYS. Scandium-46 is used in the petroleum industry as a tracer in refinery crackers, which break down large hydrocarbon molecules into more useful, smaller ones. The Stasi, the secret police of former East Germany, used it to track people with a chemical marking substance called "spy dust" by the CIA. This political application would have been just the kind of subject considered unfit for casual conversation by scandium's Swedish discoverer, Lars Nilson (1840–1899), of whom a colleague once said:

> Woe to him who dared speak of political or philosophical matters when Nilson intended to be merry. . . . He had a thousand devices for putting a stop to a conversation which threatened to take a tiresome turn. He

would, for example, sit listening for a while with a grave face, and then interpose with a short nonsensical observation, delivered with great solemnity in the accents of some political or scientific worthy of pedantic fame, while a gleam of fun shot forth from under his heavy, dusky brow.

Pity he died too soon to write a field guide to radiation.

Selenium-79

A RADIOISOTOPE OF THE ELEMENT selenium that is part of the troublesome junk of the nuclear world. As a FISSION product of URANIUM-235, it is found in spent nuclear fuel, fuel reprocessing waste, and FALLOUT. An emitter of BETA PARTICLES with a HALF-LIFE of about 327,000 years, it is one of seven long-lived fission products (including CESIUM-135, IODINE-129, palladium-107, TECHNETIUM-99, tin-126, and ZIRCONIUM-93) that menace the outlook for safely disposing of the enormous quantity of reactor waste piling up at nuclear power plants around the world. These monsters are not highly radioactive, but they need to be isolated somehow from the biosphere for millennia, a task that currently has no workable solution and seems, on the face of it, absurd.

Garlic is especially high in selenium, a mineral that is essential for good health, so some way will have to be found to prevent it from being planted near waste depositories ten thousand years from now, assuming the popularity of Italian food continues. Right, then—any suggestions, class?

Sievert

THIS IS ONE of the more subtle members of the motley collection of radiation measurement units used by scientists, physicians, and engineers. It was named in 1979 after Rolf Sievert (1896–1966), a Swedish physicist who performed seminal studies of the biological effects of radiation, especially for medical diagnosis and therapy. He pioneered the standardization of dosage levels in radiation treatments, a badly needed step during an era that began with primitive notions such as the "skin erythema dose" and the "tolerance dose." Sweden's first radiation protection law arrived in 1941 thanks to his insistence.

The sievert (abbreviated Sv) was introduced specifically to help protect people and is arguably the most crucial unit for laymen to understand. Though it is expressed in the same physical terms as the GRAY—joules of absorbed radiation per kilogram of matter—it is expressly designed for *living* matter in order to reflect the fact that there are different kinds of radiation and different kinds of biological tissue.

The sievert is used in two important measurements. The EQUIVALENT DOSE, widely encountered in literature about the health effects of radiation, is the average dose absorbed by an organ or tissue, which is expressed in grays, multiplied by a quality factor that weights the influence of the type of radiation absorbed. For example, ALPHA PARTICLES are assigned the highest weighting factor, 20, because they are the most damaging to living organisms. Photons, which comprise GAMMA RAYS, are valued the lowest, at 1, as are BETA PARTICLES. NEUTRONS are rated 5, 10, or 20, depending on how energetic they are. Thus, the equivalent dose for gamma and beta radiation, in sieverts, is the same as the absorbed dose in grays. The equivalent dose of alpha particles, however, is twenty times the absorbed dose in grays. And neutrons can be five, ten, or twenty times higher.

A second measurement, called the EFFECTIVE DOSE, takes into consideration not only the type of radiation absorbed, but also the different levels of damage done by the

Rolf Sievert in his Radiumhemmet (Sweden's first radiation oncology clinic) laboratory, Stockholm, 1929

radiation to different kinds of tissue. Weighting factors derived from laboratory findings, models, and epidemiologic evidence are assigned to various tissues and organs. When the equivalent dose to a certain tissue or organ is multiplied by one of these factors, the result is the effective dose, also expressed in sieverts. This is the best measure for how dangerous—or effective in killing malignant cells, if the target is a cancerous tumor—any particular exposure to the body might be. Many parts of the body have been rated, ranging from the skin, at 0.01; to the thyroid, at 0.05; to the lungs, at 0.12; to the gonads, at 0.2. The higher the rating, the more sensitive the tissue. These values are derived from a reference population consisting of both sexes and widely varying ages. From time to time they are revised as new data become available. For whole-body exposure to a certain type of radiation, the effective dose equals the equivalent dose.

The sievert has largely superseded an older unit called the rem (for "Roentgen equivalent man"). The rem also made use of weighting factors and was based on a unit of radiation exposure called the roentgen, named in 1928 after German physicist Wilhelm Roentgen (1845–1923), who discovered X-RAYS. One sievert equals 100 rem.

Unfortunately, these important technicalities are seldom evident in mainstream press coverage of disasters that involve radiation exposure. It's just too complicated for most newspapers and television shows. Even in the professional literature, the units may be explained differently and are not always used stringently. "Radiation units can be confusing," laments the U.S. National Research Council in one of its major reports. They may be further complicated by the appendage of time, volume, and *per caput* factors, making them confusing for specialists and nonspecialists alike. There is often no clue in official announcements, let alone in subsequent news stories, of how doses were calculated. In order to protect themselves, consumers of the news need to educate themselves, or else fall prey to the vagaries— intentional or not—of public information.

A single full-body CT scan results in an effective dose of 12 mSv. A typical mammogram, at 0.13 mSv, delivers about a hundred times less. A chest X-ray, at 0.08 mSv, still less. Excess cancers—that is, cases above the number normally expected in a population—have been found among Japanese atomic bomb survivors at dose levels of about 100 to 4,000 mSv. As always, fetuses are especially vulnerable to radiation, with excess cancers found at in utero doses as low as 10 mSv.

About one person out of one hundred will develop cancer from a single dose of 0.1 Sv above BACKGROUND RADIATION.

Silver-108m, -110m

TWO RADIOISOTOPES OF THE ELEMENT silver that turn up in the detritus from nuclear reactor accidents and atmospheric weapons tests. They are formed when natural silver in control rod alloys or other hardware is bombarded with NEUTRONS. Silver-110m, a BETA PARTICLE emitter with a HALF-LIFE of 250 days, was detected in Akamoku seaweed (*Sargassum horneri*) collected from the Japanese port of Hisanohama, thirty kilometers south of the Fukushima nuclear power plant, at 6.9 BECQUERELS per kilogram in August 2011. Silver-108m from global FALLOUT, also a beta emitter, with a half-life of 418 years, was found in marine organisms across the North Pacific during the 1960s.

As part of the pop mania about anything radioactive in the decades after World War II, hundreds of thousands of dimes—which were 90 percent silver until 1964—were irradiated at the American Museum of Atomic Energy in Oak Ridge, Tennessee, as souvenirs. A dime irradiator also appeared at the 1964 World's Fair in New York. Fun.

Sodium-22, -24

SODIUM-22, A RADIOISOTOPE of one of the crucial elements required by all animal species, has a 2.6-year HALF-LIFE and is used as a radioactive tracer for natural sodium. Sodium-24 (fifteen-hour half-life) is also used as a tracer in studies of body electrolytes and to find pipe leaks in industrial settings. Both are BETA PARTICLE emitters. Both are cosmogenic nuclides—that is, produced by COSMIC RAY collisions with argon in the earth's atmosphere—but they can also be man-made. The annual EFFECTIVE DOSE from cosmogenic sodium-22 is about 0.15μSv.

Sodium-24 is of potential concern in the operation of so-called fast neutron reactors (FNRs), which are cooled with molten sodium—sodium melts at 208°F and boils at 1621°F, which is higher than the reactor's operating temperature—instead of water. Because some sodium-24 is formed when the coolant is irradiated by neutrons from the core, these reactors must have a second, independent heat-transfer circuit away from the core to prevent radioactive sodium from leaking into the environment. This is not the only concern—molten sodium burns in contact with air and reacts with water to release explosive hydrogen. Why, one might wonder, does such a devilish device exist? Not because it produces economical power, but because it burns uranium more efficiently than conventional reactors. As long as uranium is plentiful, however, the FNR remains commercially uncompetitive.

In December 1995 the Monju demonstration FNR at Tsuruga, Japan, suffered a leak of some 640 kilograms of sodium coolant from its secondary loop, causing an intense fire that shut down the plant. It did not start up again until May 2010, and its future is in doubt as Japan reevaluates nuclear power. An FNR at the Beloyarsk power station in Russia is one of the largest in the world and the focus of international study, but it has been repeatedly plagued by sodium leaks, fires, and shutdowns. Japan reportedly paid $1 billion for the technical documentation on the reactor.

Strontium-89, -90

THERE ARE DOZENS OF RADIOACTIVE isotopes of the element strontium, but only two are of practical importance: strontium-89, an artificial creation used in cancer therapy, and strontium-90, a hazardous FISSION product formed in nuclear reactors and atomic bomb explosions. The human body treats strontium as though it were calcium, accumulating it in bones and marrow, which means its RADIOISOTOPES are both potentially useful in carefully monitored medical doses and deadly agents of leukemia and bone cancer in uncontrolled exposures.

After the Fukushima nuclear reactor disaster in March 2011, strontium-90 was found in seawater, soil, and plants at levels hundreds of times higher than remnants of FALLOUT from the cold war era of atmospheric atomic weapons tests. Strontium-89 and strontium-90 were found in forty-five of one hundred soil samples taken in June and July, in one case near the boundary of the evacuation zone at 500 BECQUERELS per square meter of strontium-89 and 130 Bq/m^2 of strontium-90. As the uncooled reactor cores heated up above

4,000°F, easily evaporated DECAY PRODUCTS of atomic fission, such as IODINE-131 and CESIUM-137, escaped first, followed by less volatile products such as strontium-90 and PLUTONIUM-239. The presence of elevated strontium at locations scores of kilometers from the reactor facility was an early clue to the severity of the accident. In the 1986 Chernobyl reactor explosion and fire, about 10 quadrillion becquerels (peta-becquerels, PBq) of strontium-90 and 115 PBq of strontium-89 were released into the environment, especially across Eastern Europe. About 622 PBq of strontium-90 and 117,000 PBq of strontium-89 were released globally by above-ground weapons tests between 1945 and 1980. With a HALF-LIFE of twenty-nine years, strontium-90 is a more insidious threat than strontium-89, which disappears by half in just 50.5 days. Both are potent emitters of BETA PARTICLES.

The U.S. Food and Drug Administration sets a limit of 160 becquerels per kilogram for strontium-90 in all components of the human diet.

Like POLONIUM-210, strontium-90 radiates considerable heat energy and has been used to power small thermoelectric generators in remote locations. Many lighthouses in the former Soviet Union relied on such devices, which, after the dissolution of that government, raised fears about theft of unguarded units by terrorists who might pack strontium-90 into so-called dirty bombs—conventional explosives that would blast radioactive material into the air just to contaminate surrounding areas. The fear has proved unfounded, so far. If the scenario ever comes to pass, we may wish to revert to living beneath earthen mounds like the Sidhe of Gaelic mythology, after which the village of Strontian in western Scotland was named, lending its name in turn to a locally mined ore from which the element strontium was isolated in 1808.

Sulfur-35

THE RADIOISOTOPE SULFUR-35, an emitter of BETA PARTICLES with a HALF-LIFE of eighty-eight days, can be formed when the stable chlorine-35 isotope found naturally in seawater salt is bombarded with NEUTRONS, which explains why it was detected in sulfur oxide aerosols soon after the Fukushima reactor disaster. Created as ocean water was dumped and pumped onto damaged cores in a desperate attempt to cool them, sulfur-35 reacted with oxygen in the air to form radioactive compounds that within weeks were picked up by instruments as far away as Southern California. Sulfur-35 is also formed naturally by the interaction of COSMIC RAYS with argon in the earth's atmosphere.

By the way, nuclear weapons tests in the Pacific Ocean during the 1950s irradiated chlorine in sea salt, which in that case produced large amounts of radioactive chlorine-36 (half-life: three hundred thousand years).

Technetium-99m

THE WORKHORSE RADIOISOTOPE used in more than three-quarters of all medical diagnostic imaging exams. The *m* stands for *metastable*, which means it quickly decays into a longer-lived isotope (in this case, technetium-99). Technetium-99m's brief HALF-LIFE of six hours and pure GAMMA radiation with no associated ALPHA or BETA PARTICLES make it desirable for human subjects.

When used as a radioactive tracer, technetium-99m is first bonded chemically to a drug that will head for specific organs that doctors need to examine. The solution is then injected into the patient, and a picture is later taken with a radiation-sensitive camera. This technique helps find cancer cells that have spread from primary tumors, for example, and there are numerous technetium-99m kits for making dozens of radiopharmaceuticals to analyze the brain, kidney, heart, skeleton, liver, and lung. The common "technetium stress test" is used to diagnose many heart problems.

Technetium-99, 99m's evil relative, with a half-life of 211,000 years, is a FISSION product of URANIUM-235 and PLUTONIUM-239 and therefore a nasty component in spent nuclear reactor fuel, and in high-level radioactive wastes from processing that fuel. It is very soluble in seawater, and liquid discharges from the Sellafield reprocessing plant in northern England (formerly called Windscale, site of a catastrophic reactor fire in 1957) into the Irish Sea have been found to contaminate regional seafood such as lobsters. It also pollutes soil from FALLOUT produced by cold war–era atmospheric nuclear weapons tests, which put about 160 teraBECQUERELS (160×10^{12} Bq) into the biosphere. The 1986 Chernobyl reactor disaster released about 0.75 TBq. It has the vexing ability to migrate though the ground into water, thereby despoiling the land around old weapons production facilities.

On March 30, 2011, nearly three weeks after the Fukushima Daiichi plant was damaged by a tsunami, technetium-99m presumably from damaged nuclear fuel rods was detected in water in the Number 2 reactor building at 16,000 Bq/cm^3.

Tellurium-129, -132

THESE RADIOISOTOPES OF THE ELEMENT tellurium, which is used in steel alloys, are FISSION products. If the alloys are part of nuclear reactor hardware, they are sitting ducks for NEUTRONS from the core. Not a problem under normal operation, they become part of the storm of radioactive detritus released into the environment by a reactor breach. Thus, the BETA PARTICLE emitter tellurium-132 (HALF-LIFE about three days) was detected in Namie-machi, about twenty kilometers from the Fukushima Daiichi plant, and other locations within one day after the March 11, 2011, disaster. Concentrations ranged from 23 to 113 BECQUERELS per cubic meter of air, exceeding the regulatory "safe" limit of 20 Bq/m^3. A week later, traces of tellurium-132 were detected on the other side of the Pacific Ocean, in Seattle, Washington. Tellurium-129 (half-life about seventy minutes) and tellurium-129m (half-life about thirty-four days) were also detected in the Kanto region of Japan during March. TEPCO, the utility company that operated the plant, offered this explanation: "As the pressure inside the [Reactor No. 1] containment vessel rose, tellurium, along with hydrogen, may have escaped from the joints [on the containment vessel]. The pressure inside the reactor building also rose, and then tellurium leaked outside the building and was carried by the wind and spread wide." Thus does a piece of steel circle the globe.

Tellurium has no biological role and is highly poisonous above and beyond being radioactive. People who breathe as little as $10\mu g/m^3$ in air may develop telltale "tellurium breath" from the body's conversion of excess tellurium into volatile dimethyl telluride, which smells like garlic. There were no reports of halitosis from Fukushima, but folks had other problems on their minds.

Thallium-201

THALLIUM "STRESS TESTS" that image the flow of blood through vessels into the heart use the tracer thallium-201, a RADIOISOTOPE with a HALF-LIFE of about three days. It could take up to a month for the radiation (X-RAYS and GAMMA RAYS) from a medical diagnostic injection of thallous chloride to fall to levels that are no longer measurable above natural BACKGROUND RADIATION, but finding out the cause of chest pains or shortness of breath is worth the trouble.

Thorium-231, -232

THORIUM IS THREE TIMES more abundant in nature than uranium and occurs almost entirely as the long-lived RADIOISOTOPE thorium-232 (HALF-LIFE 1.4×10^{10} years), which is called "fertile" in the sense that it can absorb NEUTRONS and transmute into fissionable URANIUM-233. Since the birth of nuclear power technology, the potential of thorium-232 for breeding man-made uranium-233 in a reactor has been a carrot at the end of a very long stick for proponents of this energy source. India, which possesses large thorium deposits and has a long-term goal of becoming energy-independent based on this resource, is especially eager to develop such reactors, which are not yet commercially viable. On the other hand, Australia also holds vast thorium reserves but is opposed to the development of a nuclear power industry, with or without thorium-based systems.

The World Nuclear Association, an international lobby for nuclear energy, maintains that uranium will not soon be displaced as a commercial reactor fuel. "The uranium fuel cycle is 50 years old, world proven, thoroughly mature and the costs are well known," Ian Hore-Lacy, a spokesman for the group, told *The New York Times* in 2009. Environmental advocates also regard thorium-based reactors as beyond the far horizon. "While thorium partially addresses some of the downsides of current commercial reactors based on uranium fuel," Jan Beránek, nuclear energy project leader of Greenpeace International, said, "from what we know it still has significant issues related to fuel mining and fabrication, reactor safety, production of dangerous waste, and hazards of proliferation."

Thorium-231 is the first DECAY PRODUCT of good old URANIUM-235, with a half-life of 25.5 hours.

Tritium

THE ELEMENT HYDROGEN has one RADIOISOTOPE, called tritium (or hydrogen-3, because its nucleus contains two neutrons in addition to ordinary hydrogen's single proton). It is a weak emitter of BETA PARTICLES with a HALF-LIFE of 12.3 years. Minute quantities are produced naturally by the interaction of COSMIC RAYS with nitrogen and oxygen in the earth's atmosphere, where it forms so-called tritiated water (HTO, rather than H_2O)—practically indistinguishable from normal water—that falls down in rain. The steady supply of cosmogenic tritium around the planet is only about 7 kilograms. The oceans contain about 100 BECQUERELS per cubic meter and inland water about 400 Bq/m^3, which results in humans ingesting an annual average of about 500 Bq, which gives a radiation dose of 0.01 MICROSIEVERTS.

If this were the whole story, tritium would be trivial. But after its discovery by the great Ernest Rutherford and colleagues in 1934, it found a star role in the military-industrial world of nuclear energy. Led by the dark prince of atomic weapon designers, Edward Teller (1908–2003), American scientists and engineers invented the thermonuclear (FUSION)

H-bomb in the early 1950s, wherein tritium is fused with another hydrogen isotope, called deuterium (or hydrogen-2, aka "heavy hydrogen," because its nucleus contains one neutron and one proton), to produce an enormous burst of energy—about seventeen times more than comes from splitting uranium or plutonium by FISSION, such as in the Hiroshima and Nagasaki bombs.

An early 1-megaton H-bomb contained a few kilograms of the compound lithium deuteride that was bombarded with neutrons to produce free deuterium and tritium, plus several kilograms of plutonium that fissioned to create the billion-degree-centigrade heat required for fusion reactions, and about 100 kilograms of URANIUM-238 that blanketed the whole works and fissioned for even more bang. Thus, a weapon that weighs a few hundred kilograms at most—light enough to put on the nose of an Inter-Continental Ballistic Missile (the Nagasaki plutonium fission bomb, by comparison, weighed about 4 tons)—releases as much energy as a billion kilograms of TNT. Here was a Faustian bargain that no great nation could refuse.

The governments of the United States and the Soviet Union went into overdrive to produce tritium for H-bombs. In specially designed nuclear reactors, lithium compounds enriched in lithium-6 were irradiated with neutrons to create lithium-7, which splits to form tritium and helium. During the cold war, the United States accumulated some 225 kilograms of pure tritium in this fashion and it is reasonable to assume that the Soviets did, too. The production plants

became notorious for leaking tritium into the environment—
it is, after all, just like water. Exposure to tritium among
nuclear workers has been associated with increased risk of
prostate cancer.

Tritium is also spawned as a fission product in nuclear
weapons tests, whose contribution to the atmosphere—some
240 EBq—peaked in 1963, and in nuclear power reactors,
especially those cooled and moderated by "heavy water" (deuterium oxide, D_2O). Tritium-contaminated groundwater
exists as a legacy of underground testing at old test sites in
Nevada. A large commercial nuclear power reactor will produce about 2 grams of tritium a year, which is generally
found in the fuel and its cladding. Of the sixty-five
operational commercial nuclear power plant sites in the
United States, thirty-eight have had leaks or spills—often
from corroded buried piping—that involved tritium at concentrations above the Environmental Protection Agency's
20,000-picoCURIES-per-liter threshold for safe drinking
water. (None is known to have reached public water supplies.)
One of the highest-known tritium readings was discovered
in 2002 at the Salem nuclear plant in New Jersey. Tritium
leaks from a spent fuel storage pool polluted groundwater
under the facility—located on a Delaware Bay island—at a
concentration of 15 million picocuries per liter. According
to Nuclear Regulatory Commission (NRC) records, the tritium readings in 2010–11 were still as high as 1.1 million
picocuries per liter.

Phosphorescent paints containing tritium have replaced

radium-laced materials that were banned around the world decades ago. Self-luminous signs, wristwatches, dials, and gun sights may leak tritium, but its weak beta radiation cannot penetrate skin. Tritium has been detected in landfills, however, probably from Emergency Exit signs in construction/demolition waste, which can contain 25 curies of tritium when new. In a 2010 study by the state of Pennsylvania, virtually all of its fifty-four active landfills contained tritium above natural background levels. The NRC requires a license to manufacture or transfer any product containing tritium.

Uranium-234, -235, -238

I F NOT QUITE the father of the twentieth century, uranium is at least a candidate for chairman of the board. When Henri BECQUEREL noticed on a cloudy Paris day in 1896 that a photographic plate had been fogged by something *inside* a closed drawer, the culprit was uranium in a salt compound, potassium uranyl sulfate [$K_2UO_2(SO_4)_2$]. Two low-ranking colleagues named CURIE soon stole his thunder by performing the Herculean labor of refining tons of uranium-laced pitchblende ore by hand to obtain its radioactive constituents. From there, the road led to nuclear reactors, atomic bombs, and some scary ways either to create or to kill cancer cells.

Uranium is a ubiquitous component of the earth's soil and water, identified as an element in 1789 by the German chemist Martin Klaproth (1743–1817), who named it after the recently discovered planet Uranus. Pitchblende was his muse, too, taken from the Ore Mountains of Saxony, where miners had always been intrigued by its great density but had not known quite what to do with it. Klaproth's successors in the Manhattan Project of World War II figured out one very compelling thing to do with it, and for the next decades, uranium was the most coveted substance on the planet.

In nature, uranium exists as three unstable RADIOISO-TOPES that emit ALPHA PARTICLES: mostly uranium-238 (about 99.3 percent), uranium-235 (about 0.7 percent), and trace amounts of uranium-234. They are around because each has a very long HALF-LIFE (about 4.5 billion years, 700 million years, and 250,000 years, respectively). There are a score of others that can be created artificially, but most have such exceedingly short half-lives that they are useless. Uranium-235 is highly valued because it will engage in FIS-SION reactions that release vast amounts of energy—so much that expending enormous energy to produce ENRICHED URANIUM that contains higher percentages of it is considered well worth the effort. A single kilogram of uranium-235 can theoretically produce as much energy as burning 3 million kilograms of coal. In the "Little Boy" bomb dropped on Hiroshima in 1945, less than a kilogram of uranium-235 caused an explosion equivalent to at least 12.5 thousand tons of TNT. Uranium-238 is prized because it can be used to generate the dark prince of all radioiso-topes, PLUTONIUM-239, for bombs and reactors. Even the material left over after enrichment, which is almost pure uranium-238 (called depleted uranium) is readily utilized in shielding and armor-piercing ammunition, where its high density is desirable. Uranium's status as a magical key to international political power after World War II resulted in experiments on humans that epitomized the cold war era. During 1946 and 1947, while the Nuremberg Trials were publicizing the atrocities of Nazi doctors, the Manhattan Project funded uranium toxicity studies at the University of Rochester, using hospital patients as subjects without their informed consent. The goal was to find the minimum dose that would cause kidney damage. Subjects included four men and two women, all with healthy kidneys, ranging from twenty-four to sixty-one years old. One patient was diagnosed as being in a "hallucinatory state" from alcoholism; another was suffering from "emotional maladjustment"; and a third, admitted to the hospital for a fifth time, was noted as follows: "As he had no home, he agreed willingly to enter the metabolic unit for special studies." Highly enriched uranium (that is, containing enhanced quantities of uranium-234 and uranium-235) was administered intravenously in increasing doses until there were signs of kidney malfunction. Afterward, there was no follow-up to monitor the health of the subjects. In 1986 the study was cited in a report from the U.S. House Committee on Energy and Commerce, titled "American Nuclear Guinea Pigs: Three Decades of Radiation Experiments on U.S. Citizens."*

* See also S. H. Bassett, A. Frankel, N. Cedars, H. Van Alstine, C. Waterhouse, and K. Cusson, "The Excretion of Hexavalent Uranium Following Intravenous Administration. II. Studies on Human Subjects," University of Rochester, UR37, June 1948.

Xenon-133, -135

AFTER THE FUKUSHIMA Daiichi disaster, the RADIO-ISOTOPE xenon-133 was detected at levels as high as 20 exaBECQUERELS (20×10^{18} Bq)—the largest radioactive noble gas release not associated with FALLOUT from cold war nuclear weapons tests. Some 6.5 EBq (6.5×10^{18} Bq) was released by the 1986 Chernobyl disaster. Xenon-133 and -135 are FISSION products of URANIUM-235 and PLUTONIUM-239. They are both emitters of BETA PARTICLES, though with rather short half-lives: 5.25 days and about 9 hours, respectively.

Xenon-133 and xenon-135 have often been detected after underground weapons tests and are used in the forensic work of monitoring for clandestine explosions. Xenon is chemically unreactive and therefore does not coat the surfaces of cracks in rock or get washed out of the atmosphere by rain. Because the ratio of the production of these isotopes in plutonium fission is known, and because xenon-135 decays much faster, the ratio of their concentrations in a FALLOUT plume provides a rough measure of the number of xenon-135 half-lives and thus of the time elapsed since the explosion. About 60 grams of plutonium are needed to produce a bomb yield of 1 kiloton. That much fission would produce about 2 grams each of xenon-133 and xenon-135. Because of their radioactivity, these xenon isotopes can be detected at levels of merely 1,000 and 100 atoms per cubic meter of air, respectively. Attention, Iran and North Korea!

X-rays

"**I**F THE HAND BE HELD between the discharge-tube and the screen, the darker shadow of the bones is seen within the slightly dark shadow-image of the hand itself." Thus did German physics professor Wilhelm Conrad Roentgen (1845–1923) calmly describe what would become one of the great innovations of the twentieth century. In 1895 his accidental discovery of what he dubbed *X-Strahlen* (*X* for unknown) ushered in a scientific revolution as momentous as the advent of Newtonian physics in the seventeenth century. For this he won the first Nobel Prize for physics, in 1901. By then, X-ray photographs were already a public sensation that would not be matched until the revelation of atomic bombs in 1945. The hand belonged to Frau Anna Roentgen, who remarked, "I have seen my death." Evidently it was not a romantic experience.

Like GAMMA RAYS, X-rays are composed of photons—mass-less packets of energy that travel in electromagnetic waves, just like visible light—not particles. They are less energetic (that is, of lower frequency and longer wavelength) than gamma rays, but there is no official dividing line between the two. Rather, X-rays are defined as originating in the shell of electrons around an atomic nucleus, whereas gamma rays come from the nucleus itself. Higher-density materials, such as bones, absorb X-rays more than soft tissues such as skin and muscle, leaving a dark shadow on photographic film.

In the first few years after X-rays were discovered, anecdotal evidence—skin burns, hair loss—quickly arose that they could be harmful. In 1903, Thomas Edison dropped his plan to build commercial X-ray machines known as fluoroscopes after his assistant died from cancer that spread from his hands after regularly testing X-ray tubes on them. General protective measures were not adopted until 1915, in Great Britain, but X-rays continued to be abused regardless of such guidelines until the danger of even low-dose exposure became widely appreciated fifty years later.

Indisputably on a par with antibiotics as a historic lifesaving development, X-ray technology is nonetheless shadowed by myriad horror stories since it was first used to locate bullets and broken bones. For example, following Israel's independence in 1948, more than ten thousand North African immigrant children were scorched with X-rays to cure ringworm of the scalp—hardly a life-threatening condition—causing a sixfold increase in thyroid cancers. X-rays were routinely used to treat acne as well, leaving patients vulnerable to untold complications later in life.

Most human exposure to X-rays still comes from medical procedures. Diagnostic exposures to patients—such as dental and chest X-rays, mammography—in the range 0.1 to 10 mSv (see EFFECTIVE DOSE) are regulated, at least in theory, to be just enough to provide the needed clinical information, though gross errors are not uncommon. The increasingly popular computed tomography, or CAT, scan resides in the upper end of this range and beyond, to 15 mSv.

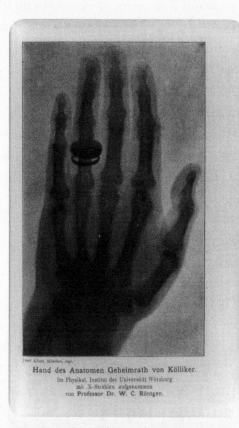

Hand des Anatomen Geheimrath von Kölliker.
Im Physikal. Institut der Universität Würzburg
mit X-Strahlen aufgenommen
von Professor Dr. W. C. Röntgen.

Frau Roentgen's lovely hand, which would have become far less so if she had left it there

Therapeutic exposures involve prodigious doses, typically in the range of 20 to 60 Gy, targeted at tumors in an attempt to kill malignant cells and halt their spread. Such powerful beams may leave patients with considerable risk of future, secondary malignancies. All X-ray machines leak somewhat, and the beams scatter, which is why technicians routinely leave the examination or treatment room when the switch is thrown. The United States and Japan have by far the most radiation treatment centers per capita in the world, therefore

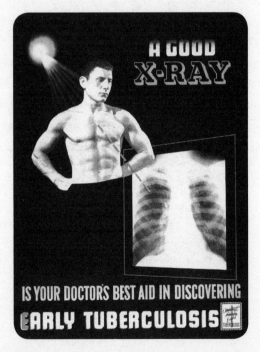

Tuberculosis prevention poster

standing out from all other countries in almost any parsing of human X-ray exposure.

So-called backscatter X-ray body scanners have been banned from airports in twenty-seven European countries, though they remain in use for security searches—including those at prisons—throughout the United States. "In order not to risk jeopardizing citizens' health and safety, only security scanners which do not use X-ray technology are added to the list of authorized methods for screening passengers at EU [European Union] airports," the European Commission announced in November 2011.

By coincidence, in November 2011 Wilhelm Roentgen finally received the honor of having a new element named after him. Roentgenium (Rg), with 111 protons in its nucleus, was created artificially in Darmstadt, Germany, in 1994 at the Society for Heavy Ion Research Laboratory. It took seventeen years for the discovery to be verified and the name approved. The longest lived RADIOISOTOPE is Rg-281, whose HALF-LIFE is twenty-six seconds.

Yttrium-90, -91

FIRST OF ALL: this element's name, which dates to 1828, comes from the Swedish village of Ytterby, where a quarry held a mineral called ytterbite. *Ytt!*

Yttrium-90 and -91 are both FISSION products that emit BETA PARTICLES, with half-lives of 64 hours and 58.5 days, respectively. Yttrium-90's significance in weapons FALLOUT and nuclear reactor waste is amplified by its troublesome parent isotope, STRONTIUM-90, which has a HALF-LIFE of twenty-nine years.

In 2002 the U.S. Food and Drug Administration approved medical injections of yttrium-90 "microspheres" for treating certain types of liver cancer. The spheres are just big enough to lodge in a tumor's vasculature, where the short range of yttrium's beta particles—about 2.5 millimeters—allows for selective killing of malignant cells. *Ytt-ytt!*

Zinc-65

A RADIOISOTOPE OF THE BIOLOGICALLY essential element zinc, used as a tracer in zinc metabolism studies. It is a BETA PARTICLE emitter with a HALF-LIFE of 245 days.

During the 1960s, zinc-65 increased dramatically in albacore tuna caught off the coasts of Oregon and Washington. The contamination was attributed to the tuna's feeding on marine organisms from the Columbia River, which was polluted with RADIONUCLIDES in effluent from the federal plutonium production facility at Hanford, Washington. River water was used as a primary coolant for the Hanford reactors, where NEUTRONS irradiated trace amounts of natural zinc-64 and other elements. Zinc-65 and CHROMIUM-51 were the most common man-made radionuclides entering the Pacific Ocean 370 miles downstream. Occupational exposure to zinc-65 among nuclear workers has been associated with increased risk for prostate cancer. Hanford is still the most contaminated nuclear site in the United States.

Zirconium-93, -95

THE ELEMENT ZIRCONIUM is used in the alloys that clad nuclear reactor fuel rods, where bombardment by NEUTRONS from the core creates the RADIOISOTOPE zirconium-93. An emitter of BETA PARTICLES with a HALF-LIFE of 1.5 million years, zirconium-93 is also an abundant FISSION product, which means that it is found in FALLOUT from old atmospheric nuclear weapons tests—some 143 EBq was dispersed globally, mostly in the Northern Hemisphere—spent nuclear fuel, and highly radioactive wastes from nuclear reactors and fuel reprocessing plants. Zirconium-95, another fission product and beta emitter, has a half-life of sixty-four days. It was found in seawater near the exhaust water outlet of reactor Number 1 at the damaged Fukushima Daiichi nuclear power plant.

The good old days: Miss Atomic Bomb 1955 at the Sands Hotel, Las Vegas. Outdoor tests of nuclear weapons were held about seventy miles away, attracting tourists encouraged by news media.

Selected Bibliography

Argonne National Laboratory (U.S.)

Radiation and Chemical Fact Sheets to Support Health Risk Analyses for Contaminated Areas, March 2007.

Congressional Research Service, Washington D.C.

Comprehensive Nuclear Test Ban Treaty: Background and Current Developments, October 5, 2011.
Fukushima Nuclear Crisis, March 15, 2011.
The Japanese Nuclear Incident: Technical Aspects, March 31, 2011.
North Korea's Nuclear Weapons, Technical Issues, February 29, 2012.
Nuclear Power Plant Sites: Maps of Seismic Hazards and Population Centers, March 29, 2011.
U.S. Spent Nuclear Fuel Storage, May 3, 2012.

Institute for Energy and Environmental Research

Radiation and Health Fact Sheet, March 17, 2011.

Institute of Nuclear Power Operators

Special Report on the Nuclear Accident at the Fukushima Daiichi Nuclear Power Station, November 2011.

SELECTED BIBLIOGRAPHY

International Atomic Energy Agency (IAEA)

Carcinogenicity of Radiofrequency Electromagnetic Fields, June 22, 2011.
Technetium-99m Radiopharmaceuticals: Manufacture of Kits; technical report series no. 466, 2008.
Thorium Fuel Cycle: Potential Benefits and Challenges, May 2005.

International Commission on Radiological Protection (ICRP)

Recommendations; ICRP Publication 103, 2007.

National Research Council (U.S.) of the National Academies

Biological Effects of Ionizing Radiation VI; *The Health Effects of Exposure to Indoor Radon*, February 19, 1998.
Biological Effects of Ionizing Radiation VII, 2006.

United Nations Scientific Committee on the Effects of Atomic Radiation (UNSEAR)

Report to the General Assembly; 1988, 1993, 2000, 2006, 2008, 2010.

U.K. Department of Health

Mobile Phones and Health, September 2005.

U.S. Air Force Institute for Operational Health

Bioenvironmental Engineer's Guide to Ionizing Radiation, October 2005.

U.S. Department of Energy

Human Radiation Experiments Associated with the U.S. Department of Energy and Its Predecessors, July 1995.

SELECTED BIBLIOGRAPHY

U.S. Department of Transportation, Office of Aviation Medicine

Galactic Cosmic Radiation Exposure of Pregnant Aircrew Members II; Federal Aviation Administration, October 2000.

U.S. Department of Veterans Affairs

Veterans and Radiation, August 2004.

U.S. Environmental Protection Agency (EPA)

A Citizen's Guide to Radon, January 2009.
Health Risks from Low-Level Environmental Exposure to Radionuclides, January 1998.

U.S. General Accountability Office

Nuclear Nonproliferation: U.S. Agencies Have Limited Ability to Account for, Monitor, and Evaluate the Security of U.S. Nuclear Material Overseas, September 2011.

World Health Organization

Guidelines for Iodine Prophylaxis Following Nuclear Accidents, 1999.
Ionizing Radiation, Part 2: Some Internally Deposited Radionuclides; IARC Monographs; vol. 78, 2001.
Ionizing Radiation, Part 1: X- and Gamma-Radiation, and Neutrons; International Agency for Research on Cancer (IARC); IARC Monographs on the Evaluation of Carcinogenic Risks to Humans; vol. 75, 2000.

Bulletin of the Atomic Scientists, "Special Issue: Low-Level Radiation Risks," May/June 2012, 68(3).

"The Human Plutonium Injection Experiments"; Moss, William, and Eckhardt, Roger; *Los Alamos Science*, no. 23, 1995.

"Identification Keys, the 'Natural Method,' and the Development of Plant Identification Manuals"; Scharf, Sara T.; *Journal of the History of Biology*, 42, 2009.

"Marguerite Perey (1909–1975): A Personal Retrospective Tribute on the 30th Anniversary of Her Death"; Adloff, Jean-Pierre, and Kauffman, George B.; *The Chemical Educator*, 10, 2005.

"Marie Sklowdowska-Curie (1867–1934): A Person and Scientist"; Mierzecki, R.; *Czechoslovak Journal of Physics*, vol. 49 S1, 1999.

"Waking a Sleeping Giant: The Tobacco Industry's Response to the Polonium-210 Issue"; Muggli, Monique E., et al.; *American Journal of Public Health*, vol. 98, no. 9, September 2008.

Index

Note: Page numbers in *italics* refer to illustrations. Page numbers in **bold** refer to chapters.

Image Credits

IMAGE CREDITS

p. 161 (Occupational radiation): From the Collection of the Public Library of Cincinatti and Hamilton County

p. 181 (Radioisotope, radionuclide): Courtesy of the National Library of Medicine

p. 185 (Radium): Courtesy of the National Library of Medicine

p. 188 (Radium): Courtesy of the National Library of Medicine

p. 192 (Radon): U.S. Geological Survey

p. 205 (Sievert): Radiumhemmets arkiv

p. 233 (X-rays): Library of Congress

p. 234 (X-rays): Courtesy of the National Library of Medicine

p. 239: University of Nevada, Las Vegas Libraries